Localizing Global Food

Short food supply chains (SFSCs) rely primarily on local production and processing practices for the provision of food and are, in principle, more sustainable in social, economic and environmental terms than supply chains where production and consumption are widely separated.

This book reviews and assesses recent initiatives on this topic from an interdisciplinary perspective. In theoretical terms it draws on and advances two key concepts, namely, place (particularly embeddedness in local economic networks and communities) and governance (particularly in addressing sustainability concerns in an inclusive and socially just manner). Empirically, the book examines a diverse set of SFSCs such as small-scale entrepreneurship, farmers' markets, community supported agriculture and grassroots and solidarity networks. The main examples discussed are from Europe and North America, but the issues are applicable in a global context.

The book is of interest to advanced students, researchers and professionals in food studies, sociology, geography, planning, politics and environmental studies.

Agni Kalfagianni is Associate Professor of Transnational Sustainability Governance, Copernicus Institute of Sustainable Development, University of Utrecht, the Netherlands.

Sophia Skordili is Professor of Industrial Geography, Department of Geography, Harokopio University, Athens, Greece.

Routledge Studies in Food, Society and the Environment

Sustainable Food Futures
Multidisciplinary Solutions
Edited by Jessica Duncan and Megan Bailey

Food Riots, Food Rights and the Politics of Provisions
Edited by Naomi Hossain and Patta Scott-Villiers

Food Sovereignty, Agroecology and Biocultural Diversity
Constructing and Contesting Knowledge
Edited by Michel Pimbert

Food and Nutrition Security in Southern African Cities
Edited by Bruce Frayne, Jonathan Crush and Cameron McCordic

The Real Cost of Cheap Food (Second Edition)
Michael Carolan

Food Bank Nations
Poverty, Corporate Charity and the Right to Food
Graham Riches

Urban Food Systems Governance and Poverty in African Cities
Edited by Jane Battersby and Vanessa Watson

Risk and Food Safety in China and Japan
Theoretical Perspectives and Empirical Insights
Edited by Louis Augustin-Jean and Jean-Pierre Poulain

Localizing Global Food
Short Food Supply Chains as Responses to Agri-Food System Challenges
Edited by Agni Kalfagianni and Sophia Skordili

For further details please visit the series page on the Routledge website: www.routledge.com/books/series/RSFSE/

Localizing Global Food

Short Food Supply Chains as Responses to Agri-Food System Challenges

Edited by Agni Kalfagianni and Sophia Skordili

Routledge
Taylor & Francis Group
LONDON AND NEW YORK

earthscan
from Routledge

First published 2019
by Routledge
2 Park Square, Milton Park, Abingdon, Oxon OX14 4RN

and by Routledge
52 Vanderbilt Avenue, New York, NY 10017

First issued in paperback 2020

Routledge is an imprint of the Taylor & Francis Group, an informa business

British Library Cataloguing-in-Publication Data
A catalogue record for this book is available from the British Library

Library of Congress Cataloging-in-Publication Data
Names: Kalfagianni, Agni, editor. | Skordili, Sophia, editor.
Title: Localizing global food : short food supply chains as responses to agri-food system challenges / edited by Agni Kalfagianni and Sophia Skordili.
Description: Abingdon, Oxon ; New York, NY : Routledge, 2019. | Includes bibliographical references and index.
Identifiers: LCCN 2018033528 (print) | LCCN 2018046276 (ebook) | ISBN 9780429449284 (eBook) | ISBN 9781138327368 (hbk) | ISBN 9780429449284 (ebk)
Subjects: LCSH: Local foods–Case studies. | Food industry and trade–Case studies. | Agricultural industries–Case studies. | Food supply–Case studies.
Classification: LCC HD9000.9.A1 (ebook) | LCC HD9000.9.A1 L63 2019 (print) |
DDC 338.1/9–dc23
LC record available at https://lccn.loc.gov/2018033528

ISBN 13: 978-0-367-58299-9 (pbk)
ISBN 13: 978-1-138-32736-8 (hbk)

Typeset in Bembo
by Wearset Ltd, Boldon, Tyne and Wear

Contents

Figures

Tables

Contributors

Theodosia Anthopoulou is Professor of Social and Rural Geography at Panteion University of Social and Political Sciences in Athens, Greece. Her recent research focuses on family farming and rural development, localized agri-food systems and alternative food networks within social economy.

Filippo Arfini is Professor in Agricultural Economics at the Department of Economics and Management, Università di Parma, Italy. His scientific activity follows three main fields of research: (a) mathematical programming models for impact assessment of Common Agricultural Policy reform; (b) agri-food product management with a special focus on food quality schemes; and (c) rural development.

Alison Blay-Palmer is Centre for International Governance and Innovation (CIGI) Chair in Sustainable Food Systems and Associate Professor in Geography and Environmental Studies, Wilfrid Laurier University at Waterloo, Canada. Her research interests focus on sustainable food systems, community well-being, city region food systems, urban agriculture, the right to food and food planning.

Claire Cerdan is Senior Researcher at CIRAD (French International Cooperation in Agricultural Research for Development) in Montpellier, France. She works on local food systems and food quality standards with a special focus on geographical indications and smallholders' market access in Southern countries.

Dimitris Goussios is Professor in the Department of Planning and Regional Development at the University of Thessaly, Greece, where he is also Head of the Laboratory of Rural Space. His research focuses on methods and tools for a territorial approach to rural/local development; territorial branding; forms of cooperation and coordination in rural areas; and the role of rural diaspora in territorial development.

Cristina Grasseni is Professor of Anthropology and Scientific Director of the Institute of Cultural Anthropology and Development Sociology at

Leiden University, the Netherlands. Her research interests lie in political, ecological and economic anthropology.

Agni Kalfagianni is Associate Professor of Transnational Sustainability Governance at the Copernicus Institute of Sustainable Development, Utrecht University, the Netherlands. Her research focuses on questions of effectiveness, legitimacy and the ethical and justice considerations of private and transnational forms of governance.

Maria Cecilia Mancini is Associate Professor in Agricultural Economics at the Department of Economics and Management, Università di Parma, Italy. She has been developing her scientific activity along two main streams: (a) the socio-economic impact of "quality labels" schemes (both voluntary and mandatory) on the actors in agri-food supply chains and territories, both in developed and less developed countries; and (b) the economics of animal welfare.

Kevin Morgan is Professor of Governance and Development in the School of Geography and Planning at Cardiff University, where he is also Dean of Engagement. Sustainable food systems is one of his main research interests and in that field he has focused on food provisioning in public places and the role of municipalities in food planning and food framing.

Phil Mount is Research Associate at the Laurier Centre for Sustainable Food Systems, Wilfrid Laurier University at Waterloo, Canada and Associate Director of Just Food in Ottawa, Canada. His research interests include sustainable regional development, agroecology and agricultural supply management programmes.

Mike Nagy is Research Associate at Wilfrid Lauriel University Centre for Sustainable Food Systems, Wilfrid Laurier University at Waterloo, Canada. He has been intimately involved in the local food and water protection movement for over 15 years and currently works in the culinary field by supplying boat-to-table Canadian fish to chefs.

Erin Nelson is Assistant Professor in the Department of Sociology and Anthropology at the University of Guelph, Ontario, Canada. Her research interests focus on sustainable food systems, food security, agroecology, community development, participatory research and Latin America.

Lena Partzsch is Professor of Environmental and Development Policy at the University of Freiburg, Germany. Her research interests lie in the fields of sustainability governance and international relations.

Coline Perrin is Researcher at INRA (the French Institute for Agricultural Research), lab UMR Innovation, Montpellier, France. Her work focuses on the influence of urban planning on farmland management and on the interconnections between cities, agriculture and food in the Mediterranean.

Birgit Peuker is Research Associate at the Institut für Sozialinnovation ISI-consult (Institution for Social Innovation), Berlin, Germany. Her research interests revolve around environmental and organizational sociology.

Raychel Santo is Senior Research Program Coordinator at the Johns Hopkins Center for a Livable Future. Her research focuses on local and regional food governance, urban agriculture, institutional food procurement and the relationship between diet and climate change.

Sophia Skordili is Professor of Industrial Geography in the Department of Geography, Harokopio University, Athens, Greece. Her research interests focus on the impact of globalized agri-food networks on small firms and rural areas; the dynamics of cooperative small and medium sized enterprises and short food supply chains in sustainable local development; and local agri-food systems' linkages with tourism.

Elodie Valette is Senior Researcher at CIRAD (French International Cooperation in Agricultural Research for Development), Montpellier, France. She works on the coexistence of different land uses at the urban–rural interface, focusing on the practices of adaptation, resistance and innovation that actors develop therein.

Preface

The idea behind this book was born through our common concerns about the way the dominant industrialized agri-food system operates and its gruelling consequences on the natural environment and human well-being. It is well documented that the way food is produced and consumed in the majority of places around the world today exceeds environmental limits, affects human health and fails to provide food security for the most vulnerable members of global societies. And yet, while there is a clear need to transform the conventional agri-food system it is unclear how this can be done successfully.

In a search for alternatives we observed a growing trend – particularly in the Global North – towards food localization initiatives ranging from farmers' markets to community supported agriculture, local food hubs and food retailers supporting local suppliers, to name a few. These alternatives aim to shorten the distance between production and consumption and promise to provide a counterbalance to the largely obscure and anonymous conditions of food production that characterize the conventional agri-food system. But do they live up to this promise?

To explore this question in detail, we invited scholars from various disciplines – geography, planning, political science, anthropology and economics – and geographical settings to provide insights on the potential and pitfalls of short food supply chain initiatives to foster a more sustainable agri-food system. In October 2016, we organized a two-day workshop in Utrecht to bring together the contributors of this book. During the two-day workshop the authors presented the first draft of their chapters and we had the opportunity to define the main aims and the distinctive features of the book. The result is a rich compilation of perspectives, approaches and empirical cases that can inform and nuance the growing literature on food localization and short food supply chains.

We would like to thank all the contributors for their hard work as well as their patience and diligence in developing and revising their individual chapters. We would also like to thank Tim Hardman, the book series editor, for supporting our book, Amy Johnston for her excellent editorial assistance and two anonymous reviewers for valuable comments and criticism. A special thanks goes to the Institutions for Open Societies at Utrecht University for providing funding for the authors' workshop that brought us all together in the crucial first stages of the book's development.

Agni Kalfagianni and Sophia Skordili

Abbreviations

AFN	alternative food networks
AMAP	*Association pour le Maintien d'une Agriculture Paysanne*
C$	Canadian dollars
CAP	Common Agricultural Policy
CDO	Controlled Designation of Origin
CFN	conventional food network
CFPR	Parmigiano-Regianno Consortium (English translation)
CFS	conventional food system
CFSC	Community Food Security Network
CIRAD	French Agricultural Research Centre for International Development
CLF	(Johns Hopkins) Center for a Livable Future
CSA	community supported agriculture
CSO	civil society organization
DES	*distrettidi economia solidale* (districts of solidarity economies)
EU	European Union
FAO	Food and Agriculture Organization (FAO) of the United Nations
FARA	Food Access Research Atlas
FCWA	Food Chain Workers Alliance
FIAN	FoodFirst Information and Action Network
FMPP	Farmers' Market Promotion Program
FPC	food policy council
FPN	food policy network
GAS	*Gruppi di acquisto solidale* ("solidarity purchase group")
GF	Good Food
GHG	greenhouse gas(es)
GI	Geographical Indication
GIS	Geographic Information System
GM	genetically modified
GMO	genetically modified organism
GPS	Global Positioning System
HACCP	Hazard Analysis and Critical Control Points

ICT	information and communications technology
IFOAM	International Federation of Organic Agriculture Movements
INRA	French National Institute for Agricultural Research (English translation)
ISO	International Organization for Standardization
LAFS	localized agri-food system(s)
LFPG	local food policy group
LFS	local food system(s)
LSFS	local sustainable food system
MNC	multinational corporation
MUFPP	Milan Urban Food Policy Pact
NGO	non-governmental Organization
n.y.	no year
OECD	Organization for Economic Cooperation and Development
OMAFRA	Ontario Ministry of Agriculture, Food and Rural Affairs
PDO	Protected Designation of Origin
PGI	Protected Geographical Indication
PR	Parmigiano-Reggiano
RES	*reti di economia solidale* ("networks of solidarity economies")
SDG	Sustainable Development Goal
SEM	Single European Market
SFCN	Sustainable Food Cities Network
SFS	sustainable food system(s)
SFSC	short food supply chain
SSC	short supply chain
SMEs	small and medium enterprises
SYAL	*Systèmes Agroalimentaires Localisés*
TNC	transnational corporation
TPGS	Territorial Participatory Guarantee System
UN	United Nations
USDA	United States Department of Agriculture

1 Localizing global food

An introduction

Agni Kalfagianni and Sophia Skordili

Introduction

On 25 September 2015, at the seventieth anniversary of the United Nations (UN), 193 heads of state and government agreed on 17 Sustainable Development Goals and 169 targets, which will guide UN action for the next 15 years. Goal 2, Zero Hunger, sets specific targets to "end hunger and all forms of malnutrition by 2030" (United Nations, 2015) and interacts very closely with environmental, economic and social justice goals (FAO, 2016). Accordingly, food and agriculture assume centre stage in the 2030 global Agenda for Sustainable Development. Concurrently, there is increasing realization at the international level that responses to the multiple challenges facing the agri-food system need to be interconnected and systemic.

In this context, short food supply chains (SFSCs), which aim to shorten the distance between production and consumption in the provision of food (Marsden *et al.*, 2002), appear promising. Ranging from farmers' markets, to community supported agriculture (CSA) projects, and to alternative food retail chains – to name a few examples – SFSCs appear as innovative forms of reorganizing the food system from global to local, giving visibility and face to the largely obscure and anonymous conditions of food production that characterize the conventional food system (CFS).

But do they live up to this promise? While the literature has started to examine SFSCs from a variety of perspectives, this has not, as yet, been carried out in a systematic and comprehensive manner. In particular, a large proportion of monographs and edited volumes have a specific focus on the rise of social movements in this context, rather than on SFSCs. Thus, Gottlieb and Joshi (2010) recount the evolution of the food justice movement in the USA. Alkon and Agyeman (2011) investigate how race and class inequalities develop in the food system and give future directions for the food justice movement. Hayes-Conroy (2014) examines the evolution of the organic movement in the USA; studying school gardens and slow food movements in Berkeley, California and rural Nova Scotia, she questions the tendency of alternative food activism to promote sameness over social difference. Hunt (2015) undertakes a historical analysis of the local food movement in the UK and the USA and examines, in a comparative fashion, its role in developing effective national policy strategies. Two relatively recent reports by Santini *et al.* (2013) and Galli and Brunori (2013) provide a meta-analysis of the literature on SFSCs regarding their impact on sustainability and health followed by a number of case-studies from European Union (EU) countries. Although insightful, these reports are largely empirical and lack the theoretical basis which could contextualize and explain the observed impact. Moreover, a number of edited volumes examine broader questions of the effectiveness and legitimacy of agri-food system governance but lack any specific focus on SFSCs (e.g. Clapp and Fuchs, 2009; Havinga *et al.*, 2015; Rayfuse, 2012; Spaargaren *et al.*, 2012).

There is a gap, then, in a theoretically informed and systematic understanding of the potential of SFSCs to provide an adequate response to the economic, social and environmental – i.e. sustainability – agri-food system challenges. This book aims to address this gap. Analytically, it draws on and advances two key concepts: *place* (particularly embeddedness in local networks and communities) and *governance* (particularly the aspects of agency – i.e. the capacity to act and constitute in the context of plurality – and diversity and fairness – i.e. the ability to respond to the three dimensions of crisis in an inclusive and socially just manner). Empirically, the book draws on a variety of cases from Europe and North America.

The book has three main features which together signify its distinctiveness in the broader literature of food and agriculture and alternative forms of supply chain organization. More specifically: (a) it has a very distinct and solid focus on SFSCs and their variety instead of focusing on a specific type; (b) it debates SFSCs both theoretically and empirically as a viable alternative to addressing the sustainability challenges facing the contemporary agri-food system, and not simply as a temporary solution for the short term and for limited audiences; and (c) it draws on examples from the Global North, highlighting that the sustainability challenges facing the agri-food system are not endemic to developing countries but are evident at the very heart of the prosperous capitalist world, thus also underlying the need to radically rethink the global governance of food.

The broader context: the malfunctions of the conventional food system

Sustainable development is defined as "development that meets the needs of the present without compromising the ability of future generations to meet their own needs" (World Commission on Environment and Development, 1987). It consists of three pillars which seek a balanced approach towards achieving economic and social development and environmental protection. Scholars have long recognized that the conventional food system is challenging all three dimensions of sustainability. In the following we present a short overview of these challenges. Each heading is understood in its wider sense so as to provide a fuller account.

Economic

In the economic dimension of sustainability we observe three main challenges: the escalating dominance of transnational corporations (TNCs) in food supply chains; the externalization of sustainability costs on the basis of efficiency arguments; and the price volatility of commodities as a result of the financialization and concentration of the agri-food system.

More specifically, the power and global reach of the leading agro-industrial TNCs has reached unprecedented heights. Concentration is extremely high in the technology intensive chemical, pharmaceuticals and biotechnology industries, the engines of industrial agriculture. The same six TNCs control 75 per cent of all private sector plant breeding research, 60 per cent of the commercial seed market and 76 per cent of global agrochemical sales (ETC Group, 2013). In parallel, large processing industries, trading companies and supermarket chains – the "food empires", according to van der Ploeg – increasingly exert a monopolistic power over the entire food supply chain (van der Ploeg, 2010). Private capital moves around borders and controls the global food system, under dense private regulation schemes (Kalfagianni, 2014), and is largely unaccountable. Importantly, a decreasing proportion of total value added is captured by corporate actors in long supply chains, with devastating effects on small family farms, processing firms and independent retailers.

Supporters of the large industrial agri-food model argue that, although far from ideal, food is cheaper and safer than ever before (Fresco and Poppe, 2016). They further note that for several developed countries, the proportion of household income spent on food consumption has dropped drastically (EIU, 2015). According to the neoliberal view, the success of the conventional food system is the triumph of the economic mechanisms of economies of scale and free trade. The large volume of production can guarantee low production costs while the specialization of trade partners along the lines of the principle of comparative advantage can lead to benefits for all. However, the economic efficiency of the conventional food system is highly questioned. It is well documented that the main advantage of big firms should be traced to their ability to externalize various sources of economic, social and environmental costs. The internalization of these negative externalities into prices would increase the cost of "real" food considerably with shattering effects on agri-food corporations' profitability. According to the New Economics Foundation, if adverse environmental impacts were included, the total external cost of the UK food system is estimated to be between £11 billion and £26 billion. Hence, consumers' effective food bill is at least 12–28 per cent greater than the price paid at the till (NEF, 2010: 5).

Further, the recent global food and financial crisis in 2008 revealed the inherent instability and vulnerability of the prevailing conventional food system. Price volatility for basic food commodities increased the number of hungry people by 75 million during 2008 and demonstrated the inefficiencies of international trade in turbulent times (McMichael and Schneider, 2011). As Clapp (2014) notes, such volatility is closely associated with the financialization of the global food system, i.e. the increasing role of financial investors on agricultural commodities futures markets. Likewise, the financial strand of the crisis revealed that relying only on a few gigantic food suppliers can be a very risky choice for the resilience of the food provision system. Due to the fundamental importance of uninterrupted food provision a more diversified production base is considered necessary in order to strengthen the resilience of the system against shocks (Sage, 2013).

Social

In the social dimension the conventional food system has been associated, among other things, with adverse health effects, lack of transparency and exclusionary consequences for local varieties of plants necessary for the survival of local food cultures.

First, the conventional food system has led to large scale homogenization of food production and consumption patterns. Heavily advertised mass produced global brands enjoy a large share in international markets. These are, however, criticized for the high content of sugar and saturated fat with several diet related concerns for human health. Currently 650 million people are obese and more than 1.2 billion are overweight (WHO, 2018). While obesity has attracted the focus of attention in the Global North, it is not just a problem for developed countries. As a result of the "retail revolution" (Reardon *et al.*, 2007) – the fast penetration of large grocery retail chains to the developing world – a food transition is taking place resulting in a movement away from traditional nutritional patterns towards western diet products sold by supermarkets, with associated diet and health related problems (Sage, 2013).

Second, consumers are increasingly disconnected from food production as supply chains become longer and more complex. Although a variety of food labelling and certification schemes can currently be identified, their large number and diversity often serves to overwhelm consumers rather than empower them (see Van Loo *et al.*, 2015). Simultaneously, many current efforts to increase transparency in conventional food chains are led by large TNCs which are largely unaccountable to the general public regarding their effects on (often small) suppliers and places of production.

Third, the conventional food system has been associated with the exclusion of cultural long-standing local varieties of plants and breeds which rarely comply with the strict unified quality standards set by the coordinators of long food supply products. Accordingly, they are excluded from global chains. This results in loss of taste, local variation and traditions, which is further accompanied by the loss of local knowledge and eradication of long-standing community and social networks articulated around food.

Environmental

In the environmental dimension, challenges are multiple and cumulative. Scholars warn that agriculture, in particular, is "a major force driving the environment beyond the 'planetary boundaries' ", a set of earth system thresholds that have been proposed as essential to keep the earth hospitable for the human species (Foley *et al.*, 2011). Freshwater use and biogeochemical flows are two of the boundaries particularly related to agricultural activities, but changes in biosphere integrity and climate change are closely connected as well.

More specifically, intensification of agricultural production has been associated with increased demands on water availability for irrigation (Molden and Fraiture, 2004), increased pollution from agricultural input and soil loss (Pretty *et al.*, 2005) and increased demand for energy (Matson *et al.*, 1997). Indeed, irrigated agriculture accounts for about 80 per cent of global water use (Molden *et al.*, 2007), while water scarcity is a critical concern, particularly in developing countries (Fedoroff *et al.*, 2010). Furthermore, extensive use of fertilizers pollutes both land and water, while studies simultaneously warn about a global depletion of phosphorous, which is essential for plants to grow (Gunther, 2005; Smil, 2000a, 2000b) and expected to reach its peak by 2030 (Cordell *et al.*, 2009). Likewise, land use modification for food production results in negative impacts for the functioning of ecosystems and biodiversity (Defries *et al.*, 2004; Geist *et al.*, 2005; Wood *et al.*, 2005).

Moreover, agriculture is a major contributor to climate change, accounting for 10–12 per cent of total global anthropogenic greenhouse gas (GHG) emissions in 2005 (IPCC, 2007). The Food and Agriculture Organization of the United Nations (FAO) estimates that agricultural emissions from crop and livestock production grew from 4.7 billion tonnes of carbon dioxide equivalents (CO_2 eq.) in 2001 to over 5.3 billion tonnes in 2011, a 14 per cent increase (FAO, 2014). Simultaneously, climate change impacts on agriculture are expected to generate additional challenges for food security in Sub-Saharan Africa and South Asia (Fisher *et al.*, 2012) as well as for groups with limited adaptation capacity, in particular smallholder and subsistence farmers, pastoralists and artisanal fisherfolk (IPCC, 2007).

Next to agriculture, environmental impacts occur at all stages of the supply chain. In this context, much recent attention of both public and private sector actors has been devoted to food waste – the wholesome, edible material intended for human consumption that arises at any point in the food supply chain but is instead discarded, lost, degraded or consumed by pests (FAO, 1981) – as a major issue of contemporary concern (Grolleaud, 2002; Morgan, 2009; Parfitt *et al.*, 2010; Thönissen, 2009).

Furthermore, environmental challenges are interconnected both with each other and with social challenges (e.g. food waste and food security), suggesting that systemic thinking becomes increasingly important in our ability to grasp the full range of food related implications (Ericksen, 2008).

The emergence of alternatives

In recent decades there has been an explosion of initiatives around food that challenge the above-mentioned malfunctions of the conventional food system. These are visible in large numbers all over Europe and the USA, under the umbrella term alternative food networks (AFNs), in a variety of formats and organization models. Certain types, such as on the farm sales, farmers' markets and urban agriculture, are the re-emergence of traditional ways of food production and consumption. Others, like community supported agriculture (CSA), box schemes, online sales and food hubs, are mainly new formats. Recent data shows their impressive growth, particularly in the USA, where the number of farmers' markets more than quadrupled from 1,775 to 8,144 between 1994 and 2007. Also in the USA, the annual turnover of direct-to-consumer sales more than doubled, from US$551 million to US$1.2 billion, between 1997–2007, while 3,637 CSA schemes were operating all over the country in 2009 (Pirog *et al.*, 2014:8).

AFNs challenge specific aspects of the conventional food system. In short, they are different or even countercultural to conventional food networks (CFNs) (Tregear, 2011). There is not a concise and coherent definition of AFNs. Morgan observes that AFNs are loosely defined rather by what they are not than by what they are (Morgan, 2010). As the conventional food sector is far from coherent (Holloway *et al.*, 2007; Watts *et al.*, 2007) it is not a surprise that its antipodes, AFNs, are varied too. Under the broad term AFN we can identify highly diversified initiatives in terms of motives, organization models and ethics.

Such motives and priorities can be highly varied in different geographical and socio-economic settings. Several researchers claim that, in the USA, AFNs are predominantly social and oppositional movements stemming from the strong tradition of social movements in the 1960s (Allen, 2004, 2008; Goodman, 2003; Gottlieb and Joshi, 2010). On the other hand, the issues of environmental sustainability, food safety, nutritional value and provenance are set high on the agenda of the affluent and consumer aware societies of Western and Northern Europe (Blay-Palmer, 2008; Lang and Heasman, 2004; Morgan *et al.*, 2006; Sage, 2013). This is not the case in Mediterranean Europe which is characterized by a more sophisticated culture of food and rich historical and cultural traditions of food and place (Brunori *et al.*, 2012: 19; Ilbery and Maye, 2005: 824).

The AFN concept has not escaped criticism. One strand of critique focuses on the weak empirical basis of the alternative/conventional dualism. Holloway *et al.* suggest that this dualism should be abandoned in favour of a relational approach which *can acknowledge the contingency of what is regarded as alternative at any one time and in any one place* (Holloway *et al.*, 2007: 5). Extensive research conducted in several geographical settings has shown that hybridization is the norm rather than the exception. Most producers have a tendency to "dip in and out" of conventional and alternative modes at different times (Ilbery and Maye, 2005; Sonnino and Marsden, 2006).

Another criticism revolves around the confined orientation of AFNs to a single issue or to a narrow spectrum of issues. Too often AFNs put exclusive emphasis on a single food issue, e.g. local supply, ignoring the multi-dimensional nature of food. Still, the single emphasis on local scale is not an end if it is not accompanied by equally important socio-economic and environmental arrangements (Born and Prucell, 2006: 195–196; see also Sharzer, 2012). Furthermore, scholars also voice concerns about the potential reproduction of economic exploitation and political oppression if AFNs are not simultaneously "transformative" (Levkoe, 2011).

These weaknesses have led to the emergence of the *Good Food* concept. The term was first used by Sage but came to greater attention when it was adopted by the W.K. Kellogg Foundation in the USA in the mid-2000s (Sage, 2002). Good Food (GF) denotes locally owned and controlled food systems that satisfy, at the same time, a set of core principles that sustain and promote ecological, economic, social and moral values. This multidimensional view is the great advantage of the GF concept. Most scholars agree that the four basic elements found under any GF chain should be summarized under the labels: *healthy, fair, sustainable and affordable.* In its ideal form, in order to qualify as GF a food chain should produce nutritious food in a manner that it is environmentally sustainable. Also, all actors in the supply chain should work to fair labour practices and all consumers, even those in lower income groups, should have access to the products (Pirog *et al.*, 2014).

There are notable contradictions among these goals. In fact, a large number of AFNs, mostly found in the prosperous societies of Western Europe and North America, are socially selective to well-off, white consumers in middle and higher income groups (Allen, 2010). Empirical research has shown that organic and high quality food products have a significant price premium, usually 20 per cent, or even more, higher than conventional products (Carpio and Isengildina-Massa, 2007, cited in Kneafsey *et al.*, 2013: 14). This price premium enables producers to cover the "real cost" of food. There are no easy answers as to how to avoid social exclusiveness and meet ethical, quality and environmental standards at the same time.

Short food supply chains

Short food supply chains (SFSCs), a term introduced at the beginning of the twenty-first century (Marsden *et al.*, 2002; Renting *et al.*, 2003), is a distinctive concept within the wider umbrella of AFNs. SFSCs are a "short-circuit" of the long, anonymous supply chains that characterize the industrial mode of food production. Their defining characteristic is the re-spatialization of food, i.e. the re-localization and re-socialization of food production and consumption processes.

Two key points characterize SFSCs. First, the control of the economic activity is retained locally due to the minimal number, ideally zero, of intermediaries involved. As such, the length of the supply chain is not necessarily short in physical terms but short in terms of the number of stakeholders involved (Kneafsey *et al.*, 2013). Hence there is a strong association of the chain with the tangible and intangible assets of locality. Second, authenticity and trust are mediated through personal interaction or information flows about origin, production processes and product quality, including sustainability related attributes (Renting *et al.*, 2003). Moreover, information also usually encompasses issues of identity, values and ethics relating to the producer (Galli and Brunori, 2013).

Based on the relation between producers and consumers, the literature distinguishes three broad types of SFSCs. While this distinction serves analytical purposes, in practice any business may be involved in more than one of the following categories (Galli and Brunori, 2013; Renting *et al.*, 2003; Sage, 2002):

- *Face-to-face interaction.* Consumers buy goods directly from the producer. This is a typical form of direct sales, for example, on the farm sales, farmers' markets and CSA schemes.
- *Spatial or cultural proximity.* Products are sold within the region of production and consumers become aware of the local nature of the product at the point of sale. Products can access the regional market through several market channels, most notably farmers and speciality retailers. This category also includes SFSCs that implement sales to distant places but under cultural proximity. A good example is sales made to immigrants in the locality who are well aware of the product characteristics, the producer, the wider area, and so on.
- *Extended SFSC.* Products are sold to consumers outside the region of production who may have no personal experience of that locality. These long distance networks are still "short" food supply chains since it is not the distance over which a product is transported that is critical, but the fact that it is embedded with value-laden information when it reaches the consumer – for example, printed on packaging or communicated at the point of sale. Other sources of information are formal certificates of local production (Protected Destination of Origin, Protected Geographical Indication) and stories about the products. In this respect internet portals and social media play a critical role.

SFSCs exist in various forms and business organization models. They embrace initiatives of the public and private sectors, as well as social economy and out of market schemes.

In certain settings, mainly in Northern Europe and North America, the public sector is a significant buyer of local food via local food procurement programmes for catering in schools, hospitals, care homes, prisons, etc. (Morgan and Sonino, 2010; Wiskerke, 2009). Morgan is right to observe that catering in the public sector can contribute decisively to the formation of a sustainable local food system while, at the same time, it is an effective tool for improving the welfare of the more vulnerable groups in society (Morgan,

2015). The recent dispute between Michelle Obama and the Trump administration, on the loosening of the high nutritional standards of school meals the former had launched as First Lady, illustrates the point (BBC, 2017).

Numerous small and medium size private enterprises are active in various aspects of local food systems. Speciality food retailers, artisanal and low tech food processors, hospitality and catering firms that prioritize local gastronomy and local food, online shops and box schemes are considered valuable places for the production and market channels of local food. In addition, alternative food retailers, i.e. retailers that prioritize food that is prepared or produced by non-conventional agricultural systems, are increasingly prominent in the promotion of "local" food. Furthermore, conventional supermarket chains also promote local products – with controversial evaluations in the literature. However, there is not the same unanimity regarding conventional food retailers. Some scholars recognize that the promotion of local food by conventional supermarket chains is part of coherent supermarket sustainability strategies and that the large and stable volume of sales can be beneficial for local food suppliers. On the other hand, more radical voices are hesitant and underline that these are merely "local washing" and "green washing" activities (Brunori *et al.*, 2016; Galli and Brunori, 2013).

Recently, especially in Europe and North America, there has been a blossoming of SFSC initiatives taken by oppositional producer and consumer groups. The dramatic changes in food production practices and consumption patterns, driven by globalization forces and austerity policies, have led to the emergence of a dynamic food justice movement (Gottlieb and Joshi, 2010). There is a long list of innovative projects currently underway in the social economy or outside the market areas. Some aim to cover the basic daily needs for food of vulnerable groups, others experiment with alternative forms of socio-economic organization in land cultivation and food production, while several others have prioritized the issues of environmental justice and food rights. Their prospects and potential have attracted the interest of academics and policy makers.

The literature highlights several benefits of SFSCs. They are considered to provide a valuable source of income and premium prices for small producers compared to conventional retailers and contribute to the creation of more solid local economies (Galli and Brunori, 2013; Sage, 2013; Seems, 2010). Also, they may contribute to environmental protection and counter the out-migration of the area (Wiskerke, 2009). At the same time, however, a stream of critical research has also emerged arguing that the contribution of SFSCs to sustainability can be contested (Allen, 2010; Born and Purcell, 2006; Brunori *et al.*, 2016).

The analytical lenses: place and governance

We use two broad analytical lenses in order to examine the promise of SFSCs as responses to sustainability challenges in the agri-food system.

Place

The first analytical lens is "place". Food and place are intertwined at various scales and in several ways. As food supply chains become longer, less transparent and more complex, food production and consumption are increasingly disconnected from certain localities. As a reaction to the emergent "placeless foodscapes" and the dissociation of consumers from food production and content, there is a wide consensus for the need to re-localize food supply chains (Morgan *et al.*, 2006). Local orientation and control are a perquisite of GF chains. However, the concept of local food is rather vague and can have many meanings. Moreover, the links and associations of local food with a certain locality are highly varied too.

The move to local food systems can be traced to the new social movements of the 1960s that placed particular emphasis on human scale, eco-community and participation (Allen, 2010: 296; Feagan, 2007: 26). During the last two decades they have gained a lot of attention as a counterpoint to economic concentration, environmental degradation and consumption homogenization, driven by globalized food networks.

In its stricter mode, local food is produced and consumed in the framework of demarcated Local Food Systems (LFS). Quite often the demarcation of LFS is based on physical proximity only. In this respect, a LFS is an enclosed geographical area, with tight boundaries, where all the stages of food production and consumption take place. Farmers' markets, on farm sales, CSA and public catering schemes are usually organized along the lines of LFS. Quite recently, due to the heavy environmental footprint and inefficiencies of the international food trade that became evident during the 2008–2009 global food crisis, there has been a revival of food localization. Major cities are considering the possibility of increasing their food sufficiency and plan to replace existing land uses with agriculture in their surrounding hinterlands.

The demarcation of LFS, even in purely physical distance terms, is not an easy task. Physical distance denotes the distance of transportation or "food miles" the product travels from the point of production to the point of sale. The definition of the radius of the local is a highly subjective exercise and can have very different meanings for the various actors in the supply chains and for different groups. Seems has shown that producers and retailers tend to stretch the boundaries of the local compared to the more restricted areas set by consumers (Seems, 2010). Obviously, the scope of a LFS can vary considerably in accordance with the magnitude of the served area and the nature of the agri-food product. Thus, some scholars argue that "slow food" principles, and healthy eating based on locally grown and prepared food, can only take place in small cities (Borrelli, 2015). UK based farmers' markets, as well as the Waitrose retail chain, source fresh food from an area within a 30 mile radius around towns and small cities (Galli and Brunori, 2013). The much publicized locavores movement makes use of food suppliers located at a distance less than 50–100 miles from the place of consumption. These boundaries need to be extended considerably when applied to metropolitan regions. In New York City, for instance, the Healthy Food and Healthy Lives Act introduced in 2008 defines as locally produced food that is acquired from the state of New York or from another state which travels less than 400 miles (Morgan and Sonnino, 2010).

In a rapidly globalizing world it is quite unrealistic to expect that all the stages of food production and consumption should take place within a restricted geographical area. In fact, it is getting increasingly difficult to confirm that every agricultural input and every ingredient have been sourced locally. Too often empirical studies show that seeds, or animal feed, have come from far-away places (Brunori *et al.*, 2016; Kneafsey *et al.*, 2013). Also, the notion of LFS as homogenous and coherent closed systems in full protection from the destabilizing forces of the unknown wider "other" echoes the principles of defensive localism and can be misleading (Heinrichs, 2003: 37).

It is obvious that definitions of the "local" cannot be restricted solely to physical proximity. They should instead include more complex understandings based upon social and cultural factors. What really matters is the spatial configuration of production (Seems, 2010: 111). Hence, SFSC conceptualization is useful since it has the ability to transcend the ambiguities and perplexities that are associated with the physical, geographical demarcation of local areas. SFSCs escape the rather simplistic assumption that the food system is characterized by a robust dual structure, of "conventional" and "alternative" food or "local" and "global" supply chains. Too often these forms coexist in supplementary ways. Brunori *et al.* note that there is a highly dynamic local–global continuum where actors in SFSCs, while adapting to a changing environment, establish multiple relations and animate several chain configurations (Brunori *et al.*, 2016). More importantly, SFSCs are protected by porous boundaries that permit flows within and outside the locality. Hence, they escape defensive localism and have more in common with a diversity receptive food localization that supports the view that the content and interests of the "local" are relational and open to change (Heinrichs, 2003:37). Due to these analytical advantages of SFSCs we broadly adopt this term to cover the variety of initiatives studied in this book. While some of the contributors also refer to "good food" or "alternative food" concepts they do so within the context of the re-localization and re-socialization of food, i.e. the defining features of SFSCs.

The contributions in this volume explore the dynamics and challenges faced by several types of SFSCs located in diverse areas all over Europe and North America. One of the main aims of the book is to understand, more specifically, how "place" and "locality" may provide a response to the multiple dimensions of the sustainability crisis in the agri-food system today. The relationship between the SFSC under study and place is neither singular nor as settled as some would suppose. It is investigated through two interrelated axes.

The first is identification of the common themes, as well as the differences, faced by SFSCs in different regional settings. Certain SFSCs operating in specific localities have created opportunities for employment, local development, innovation and participation. Others, with very similar characteristics, operating elsewhere, have been unable to generate such fruitful results. Among other factors, it is well documented that the historical trajectory and the current socio-economic status of an area, the qualities of the human and social capital and institutional thickness, have a decisive effect on the success or failure of SFSCs (Allen, 2010; Brunori *et al.*, 2016; Kneafsey *et al.*, 2013). It is crucial to identify the assets and qualities of the specific place that have played a key role in the success or failure of the SFSC.

The second axis is clarification of the notion of locality and its hybrids employed in the SFSC. Food system localization is not taken for granted – it is important to clarify in which respects and under which circumstances it can provide solutions to sustainability problems, while in others it cannot. For certain products and certain types of areas localization is not a choice. More importantly, localization is not always the most ethical and environmentally friendly option (Born and Purcell, 2006; Heinrichs, 2003; Morgan, 2010). The study of SFSCs should focus on the spatial configuration of all stages of SFSCs and the kind of linkages they develop with certain types of areas.

Governance

Our second analytical lens is governance – the rules and institutions that govern the production, trade and marketing of food and agriculture (Fuchs *et al.*, 2011). Governance denotes new forms of decision making that go beyond traditional hierarchical state command-and-control regulation. It implies some form of self-regulation by societal actors, public–private cooperation and multi-level interactions in trying to develop solutions for particular problems.

Within governance arrangements we pay particular attention to the role of agency which we broadly define following Arendt (1958) as the capacity to act and to constitute in the context of plurality and diversity. Important to this understanding of agency is that it is both constrained by the conditions surrounding it and capable of transforming those conditions. To be transformative in terms of fostering the common good, agency must meet two criteria: it must perform through collective action rather than an "enlargement of the private"; and it must transcend the lifespan of the agents themselves as only then can it be *public*.

The question we address in this context is whether the cases examined here create the space for agency as defined above and, if so, for whom. In seeking answers to this question we explore how agency is currently perceived in the literature on food democracy, food citizenship and food sovereignty.

The food democracy literature emphasizes the element of participation. Scholars argue that as conflicts around food and agriculture are essentially about competing values, the sustainability of the agri-food system "must be socially and politically defined" (Prugh *et al.*, 2000: 7). Likewise, Hassanein

(2003) argues that people should seek active participation in shaping the food system instead of being passive spectators. It is through collective action and the "meaningful participation of individuals" that changes can be brought about in the food system reorienting it towards the common good (Hassanein, 2008). Food democracy also creates the space for confronting the concentration of power in the food regime and questions who gets to make decisions about such an important issue as food (Hassanein, 2008).

Food citizenship similarly highlights peoples' active participation in shaping the food system, rather than accepting the system as passive consumers. It is "the practice of engaging in food-related behaviors that support, rather than threaten, the development of a democratic, socially and economically just, and environmentally sustainable food system" (Wilkins, 2005). Furthermore, food citizenship is understood as the capacity for self-governance that diminishes a sense of dependence (De Tavernier, 2012). In this context, Welsh and MacRae (1998) argue that the conventional food system has limited food citizenship in four different ways, namely: (a) corporate control; (b) limited transparency; (c) manipulation of the supermarket environment to increase sales; and (d) proliferation of convenience foods that "deskill the eater".

Likewise, food sovereignty is closely associated with food democracy and food citizenship. Food sovereignty is defined as

> the right of peoples to define their own food and agriculture; to protect and regulate domestic agricultural production and trade in order to achieve sustainable development objectives; to determine the extent to which they want to be self-reliant; to restrict the dumping of products in their markets; [...].
>
> (Via Campesina, 2003, cited in Windfuhr and Jonsén, 2005)

Pimbert (2009:5) argues that the term can be better understood "as a transformative process that seeks to recreate the democratic realm and regenerate a diversity of autonomous food systems based on equity, social justice and ecological sustainability". It is an alternative to the corporate food regime (Akram-Lodhi, 2015; Wittman, 2011), allows individuals to gain power vis-à-vis corporate actors and aims at converting food schemes into benefits for small-holders (Altieri and Toledo, 2011; Desmarais, 2002).

Furthermore, within governance arrangements we are specifically interested in the concepts of fairness and justice. Most understandings of fairness and social justice in the agri-food system emphasize access to healthy, affordable and culturally appropriate food (Alkon and Norgaard, 2009). Fairness is about freedom from want, freedom from oppression and access to equal opportunities (Allen, 2008). This scholarship further underscores the racial and economic inequalities of the food system that primarily affect migrant communities and communities of colour (Alkon and Norgaard, 2009; Sbicca, 2012).

Beyond access and distribution, fairness and justice, more specifically, is about enabling agency through solidarity and self-determination and by fostering creative discursive spaces for citizens to think about solutions appropriate for their community and ecosystem. This implies that fairness emphasizes the reflection of the situated knowledge of the vulnerable and marginalized, such as low income people and people of colour (Agyeman and McEntee, 2014; Carolan, 2006). These scholars emphasize, then, that fairness and justice have an emancipatory role that can affect material conditions through spaces that offer the possibility of deliberating alternatives.

Moreover, scholars argue that fairness and justice are about questioning the conditions that created unfairness and injustice in the first place (Agyeman and McEntee, 2014). In this context, they warn about the appropriation of the language of fairness and justice by conventional, neoliberal policies and practices. Agyeman and McEntee (2014) demonstrate such appropriation through an example from the USA: the United States Department of Agriculture (USDA) Food Access Research Atlas (FARA), a mapping programme that allows users to search for and identify areas of low income and low food access. In this programme, the state defines crucial justice terms such as "access" on the basis of proximity to markets. Thus, according to FARA, adequate food access can be achieved through improved big retail food accessibility (proximity to supermarkets). The state, therefore, reproduces the neoliberal logic that not only can citizens buy their way out of the access problem, but that the market is the most appropriate domain to provide such a solution (Agyeman and McEntee, 2014). Similar observations are made about some local food movements that define justice purely on market terms (Cadieux and Slocum, 2015). Accordingly, these scholars argue that we should also examine who benefits from the use of the term "justice" or from its hollowing out through abuse of this term.

The structure of the book

This book focuses on various SFSCs formed in diverse areas in Europe and North America. It contains 12 chapters: ten contributions, the Introduction and Conclusions. The ten chapters are, mostly, theoretically informed case studies based on recently implemented field work and statistical surveys. They consider perspectives and practices from different actors in SFSCs at various geographical scales.

The book is organized into three interrelated parts: Part I – Innovative local agri-food governance; Part II – Local agri-food systems; and Part III – Alternative agri-food market channels.

The three chapters in Part I present innovative local agri-food governance practices that have recently been introduced in several cities across the USA and Europe. The first two chapters focus on initiatives taken by municipalities and grassroots consumer networks, while the third attempts to show the crucial differences that exist between the "citizen" stock company and CSA formats.

Chapter 2, by Kevin Morgan and Raychel Santo, examines recent developments in municipal food, which resolves into the question of how to feed cities in the face of looming climate change, widening inequality and burgeoning hunger. Among its other aims, the municipal food movement strives to fashion a more sustainable food system, which involves more localized production and consumption as well as more fairly-traded food from afar. Municipal innovators within local governments, civil society groups, academia and businesses are building long-term systemic solutions in local food strategies, policies and governance practices. The opening session highlights the scalar, sectoral and social questions. Municipal food movements in the USA and the UK are then analysed in a more granular fashion. Finally, the chapter addresses some key issues for food movement politics by arguing that a multi-scalar and multi-functional perspective helps prevent municipal food movements becoming inadvertent agents of green parochialism by highlighting the need to be globally engaged as well as locally embedded.

In Chapter 3, Crisitina Grasseni turns the focus to grassroots food consumer networks that have emerged in several parts of the Global North. In the USA, CSA is largely producer driven and follows an entrepreneurial model. Evidence from France (*Association pour le Maintien de l'Agriculture Paysanne* or AMAP), Germany (Solidarity Agriculture Movement) and Italy (Solidarity Purchase Groups) shows that Europe has seen the flourishing of consumer-driven initiatives and the collaboration of consumers and producers on a local scale. Examining in more detail the Italian case, the author proposes to encompass this myriad of local initiatives within the theoretical framework of "responsible innovation", envisaging alternative food provisioning arrangements as grassroots cultural and social action that activates consumers qua citizens; this makes them more responsible for their consumption choices and innovates the logistics of short food chains by mobilizing multiple roles and forms of collaboration for both producers and consumers.

In Chapter 4, Lena Partzsch reflects upon agency and accountability in food localization drawing on the two case studies of Regionalwert AG and Luzernenhof, around the city of Freiburg in the German Black Forest. The first is a "citizen" stock company that invests in local and organic agri-food businesses. Low-dividend capital and higher-priced food products allow here for food localization. Agency and accountability are defined through the market. By contrast, Luzernenhof is organized as a CSA project and experiments with practices beyond the market. Agents' interactions are trust based and agriculture is at the mercy of charity. The aim is to show that, while food localization is often portrayed as "best practice", it also implies decisions about alternatives and this raises questions of agency and accountability.

The four chapters in Part II explore the dynamics of various types of local agri-food systems and their actual linkages with their localities. The first two chapters focus on independent agri-food micro-enterprises that operate in strikingly differentiated settings such as Ontario, Canada and greater Athens, Greece. The following two chapters focus on two specific types of SFSCs that bear Geographical Indication (GI): Parmigiano-Reggiano cheese in Parma, Italy and Feta Thessalia in Greece.

Chapter 5, by Alison Blay-Palmer, Erin Nelson, Phil Mount and Mike Nagy, explores how to bring about change towards a sustainable food system. The first part deals with the limitations of local food values chains in Ontario, Canada. Research findings show that food hubs and sustainable food systems are limited by a lack of resources including financial, infrastructure and network capacities. These challenges point to opportunities for change that are explored, in the second part, by the promising *RIP Basque Country* and *Good Food Scotland* policy initiatives. Several aspects of these multi-scalar regional policies pave the way for the formulation of a viable sustainable food system policy for Ontario, Canada.

In Chapter 6, Sophia Skordili attempts to construct a sharper notion of the actual characteristics and development potential of newly emerged agri-food small and medium enterprises in Greece. In the framework of the severe recession and massive unemployment, the question is whether these new ventures can provide a valuable source of employment to weak actors in supply chains and give a boost to local economies. The chapter, making use of material from a recently implemented research project in the wider Athens area, explores the ventures' actual capacity and profile, as well as their impact on people and places. Research results provide mixed signals. New ventures have overcome the various long-standing drawbacks of the typical traditional small firm operating in the country. However, they still suffer from certain inherent complications that put their survival at risk. Moreover, the persistent recession has added a series of new challenges and magnified the existing ones. Hence, their long-term viability and success is in question unless they can be supported by appropriate territorial policies.

In Chapter 7, Filippo Arfini and Maria-Cecilia Mancini, aim to uncover the dynamics and prospects of the Parmigiano-Reggiano GI cheese SFSC embedded in Parma localized agri-food system (LAFS). The opening section presents the main points of the broader theoretical debates on the relationships and synergies between LAFS and SFSCs. The next part presents the main features of the Parmigiano-Reggiano LAFS. It is a mixed LAFS that includes industrial production systems typical of open LAFS, as well as artisan and semi-industrial systems typical of closed LAFS. Product reputation is closely linked to the reputation of the area and there are significant interactions and synergies between institutions and coordination structures belonging to the LAFS. These unique characteristics are very efficient and have strengthened producers' resilience to the current economic crisis.

In Chapter 8, Theodosia Anthopoulou and Dimitris Goussios explore the limits of GI schemes under the national Protected Designation of Origin (PDO) regulation for Greek Feta cheese. They argue that the specification of territorial resources within the defined PDO areas strengthens several actors in the supply chain (breeders, fodder producers, local food processors, small traders, local development agencies, municipalities as managers of communal pastures). Hence, the main objective is to create Feta PDO of origin-placed specific quality at regional level in Thessaly region (central Greece), versus

Feta PDO of standard quality for the global market. The chapter explores three main areas: (i) whether the activation of territorial assets gives perspective in creating differentiated cheese of specific origin-based quality at regional level under national GI regulation – i.e. Feta PDO from small traditional dairy territories in Thessaly region, which in turn consolidates this same PDO label; (ii) how this differentiated Greek Feta with specific attributes can be qualified and consolidated on the market through networking and cooperation – i.e. the creation of a regional dairy cluster; and (iii) how this regional cluster could be shifted from a sectoral approach (*filière*) to a territorial perspective by integrating as much as possible small traditional dairy territories in Thessaly region (human, natural, cultural milieu).

The thematic focus of the Part III lies in the development of alternative market channels.

In Chapter 9, Agni Kalfagianni discusses the potential of alternative food retailers to provide fair responses to agri-food system challenges. The chapter conceptualizes justice as a multi-dimensional concept consisting of material, social, environmental and political dimensions. Focusing on alternative food retailers in the Netherlands, the chapter evaluates their performance against the four dimensions of justice and discusses the implications for a just transformation of the agri-food system.

In Chapter 10, Coline Perrin, Elodie Valette and Claire Cerdan question the impact of open-air food markets on producers and consumers, as well as the actual and potential role of local authorities in making markets fairer. Particular emphasis is given to the links between city dwellers and farmers and to social justice issues. The study focuses on Montpellier city region in southern France. The authors combine participant observation with quick surveys among food vendors and in-depth interviews with a small number of vendors and key informants. They conclude that the renewal of open-air food markets around Montpellier appears to be socially selective. Also, the impacts of the markets on sustainable food practices and on social justice for consumers and producers are linked directly to the way local authorities articulate the market as a key element of their local food strategy.

In Chapter 11, Birgit Peuker discusses whether CSA constitutes a real alternative to the conventional food system. She considers three possibilities: (i) CSA as an alternative which replaces agri-food; (ii) CSA as a supplement choice; and (iii) as a symbol of an alternative. CSA is discussed against the background of social movement research. Social movement theories interpret social movements as a protection of the place of direct social interaction, contrary to impersonal systems and network flows. From this point of view, CSA as an alternative is not a niche which will replace the agri-food system in the future but more an arena for learning and expressing sustainability. The analysis is based on experiences with CSA in Germany and makes comparisons with developments in other European countries.

Finally, Chapter 12 reflects on the potential and limits of SFSCs as a force that can transform the contemporary agri-food system to make it more sustainable.

References

Agyeman, J. and McEntee, J. (2014). Moving the Field of Food Justice Forward through the Lens of Urban Political Ecology. *Geography Compass* 8(3): 211–220.

Akram-Lodhi, A.H. (2015). Accelerating towards Food Sovereignty. *Third World Quarterly* 36(3): 563–583.

Alkon, A.H. (2008). Paradise or Pavement: The Social Constructions of the Environment in Two Urban Farmers' Markets and their Implications for Environmental Justice and Sustainability. *Local Environment* 13(3): 271–289.

Alkon, A.H. and Agyeman, J. (2011). *Cultivating Food Justice: Race, Class and Sustainability*. Cambridge, MA: MIT Press.

Alkon, A.H. and Norgaard, K.M. (2009). Breaking the Food Chains: An Investigation of Food Justice Activism. *Sociological Inquiry* 79(3): 289–305.

Allen, P. (2004). *Together at the Table: Sustainability and Sustenance in the American Agri-Foods Movement*. University Park, PA: Pennsylvania State University Press.

Allen, P. (2008). Mining for Justice in the Food system: Perceptions, Practices, and Possibilities. *Agriculture and Human Values* 25: 157–161.

Allen, P. (2010). Realizing Justice in Local Food Systems. *Cambridge Journal of Regions, Economy and Society* 3: 295–308.

Altieri, M.A. and Toledo, V.M. (2011). The Agroecological Revolution in Latin America: Rescuing Nature, Ensuring Food Sovereignty and Empowering Peasants. *Journal of Peasant Studies* 38(3): 587–612.

Arendt, H. (1958). *The Human Condition*. Chicago, IL: University of Chicago Press.

BBC (2017). Michelle Obama attacks Trump school meals policy. *BBC News*, 13 May 2017. Available at: www.bbc.com/news/world-us-canada-39906074.

Blay-Palmer, A. (2008). *From Industrial to Sustainable Food Systems*. London: Ashgate.

Born, B. and Purcell, M. (2006). Avoiding the Local Trap: Scale and Food Systems in Planning Research. *Journal of Planning, Education and Research* 26:195–207.

Borrelli, N. (2015). *Between Smart and Slow: A Short Introduction to the Food System*. Milan: Fondazione Giangiagomo Feltrinelli.

Brunori, G., Rossi, A. and Guidi, F. (2012). On the New Social Relations Around and Beyond Food. Analysing Consumers' Role and Action in Gruppi di Acquisto Solidale (Solidarity Purchasing Groups). *Sociologia Ruralis* 52(1): 1–30.

Brunori, G., Galli, F., Barjolle, D., van Broekhuizen, R., Colombo, L., Giampietro, M., Kirwan, J., Lang, T., Mathijs, E., Maye, D., de Roest, K., Rougoor, C., Schwarz, J., Schmitt, E., Smith, J., Stojanovic, Z., Tisenkopfs, T. and Touzard, J.M. (2016). Are Local Food Chains More Sustainable than Global Food Chains? Considerations for Assessment. *Sustainability* 8: 1–27.

Buttel, F.H. (1997). Some Observations on Agro-food Change and the Future of Agricultural Sustainability Movements. In: D. Goodman and M.J. Watts (eds.), *Globalising Food: Agrarian Questions and Global Restructuring*. London: Routledge, pp. 344–365.

Cadieux, K.V. and Slocum, R. (2015). What does it mean to do food justice? Paper 3, College of Liberal Arts All Faculty Scholarship, Hamline University. Available at: http://digitalcommons.hamline.edu/cla_faculty/3.

Carolan, M.S. (2006). Ecological Representation in Deliberation: The Contribution of Tactile Spaces. *Environmental Politics* 15(3): 345–361.

Cathcart, E.P. and Murray, A.M. (1939). A Note on the Percentage Loss of Calories as Waste on Ordinary Mixed Diets. *Journal of Hygiene* 39: 45–50.

Clapp, J. (2014). Financialization, Distance and Global Food Politics. *Journal of Peasant Studies* 41(5): 797–814.

Clapp, J. and Fuchs, D. (eds.) (2009). *Corporate Power in Global Agrifood Governance.* Cambridge, MA: MIT Press.

Cordell, D., Drangert, J-I. and White, S. (2009). The Story of Phosphorus: Global Food Security and Food for Thought. *Global Environmental Change* 19: 292–305.

De Tavernier, J. (2012). Food Citizenship: Is There a Duty for Responsible Consumption? *Journal of Agricultural and Environmental Ethics* 25: 895–907.

Defries, R.S., Foley, J.A. and Asner, G.P. (2004). Land-Use Choices: Balancing Human Needs and Ecosystem Function. *Frontiers in Ecology and the Environment* 2: 249–257.

Demirbas, A. (2008). Biofuels Sources, Biofuel Policy, Biofuel Economy and Global Biofuel Projections. *Energy Conversion and Management* 49(8): 2106–2116.

Desmarais, A.A. (2002). PEASANTS SPEAK – The Vía Campesina: Consolidating an International Peasant and Farm Movement. *Journal of Peasant Studies* 29(2): 91–124.

DeWeerdt, S. (2009a). Is Local Food Better? *World Watch* 22(3): 6–10.

DeWeerdt, S. (2009b). Local Food: The Economics. *World Watch* 22(4): 20–24.

DuPuis, M.E. and Goodman, D. (2005). Should We Go "Home" to Eat?: Toward a Reflexive Politics of Localism. *Journal of Rural Studies* 21: 359–371.

Edwards-Jones, G. (2010). Does Eating Local Food Reduce the Environmental Impact of Food Production and Enhance Consumer Health? *Proceedings of the Nutrition Society* 69(4): 582–591.

Edwards-Jones, G., Canals, L.M., Hounsome, N., Truninger, M., Koerber, G., Hounsome, B., Cross, P., York, E.H., Hospido, A., Plassmann, K., Harris, I.M., Edwards, R.T., Day, G.A.S., Tomos, A.D., Cowell, S.J. and Jones, D.L. (2008). Testing the Assertion that "Local Food is Best": The Challenges of an Evidence-Based Approach. *Trends in Food Science and Technology* 19: 265–274.

EIU (Economist Intelligence Unit) (2015). Food Security in Focus: Europe 2014. EIU report sponsored by Du Pont.

Ericksen, P.J. (2008). Conceptualizing Food Systems for Global Environmental Change Research. *Global Environmental Change* 18: 234–245.

ETC Group (2013). Putting the Cartel before the Horse … and Farm, Seeds, Soil, Peasants, etc. Who Will Control Agricultural Inputs? Report 111, 09/2013.

FAO (1981). Food loss prevention in perishable crops. FAO Agricultural Service Bulletin, No. 43, FAO Statistics Division.

FAO (2014). Agriculture's greenhouse gas emissions on the rise, FAO news report, 11 April 2014. Available at: www.fao.org/news/story/en/item/216137/icode/.

FAO (2016). Food and agriculture: Key to achieving the 2030 agenda for sustainable development. FAO, Rome.

Feagan, F. (2007). The Place of Food: Mapping Out the "Local" in Local Food Systems. *Progress in Human Geography* 31(1): 23–42.

Fedoroff, N.V., Battisti, D.S., Beachy, R.N., Cooper, P.J.M., Fischhoff, D.A., Hodges, C.N., Knauf, V.C., Lobell, D., Mazur, B.J., Molden, D., Reynolds, M.P., Ronald, P.C., Rosegrant, M.W., Sanchez, P.A., Vonshak, A. and Zhu, J.-K. (2010). Radically Rethinking Agriculture for the 21st Century. *Science* 327(5967): 833–834.

Fischer, G., Shah, M. and Velthuizen, H.V. (2002). Climate change and agricultural vulnerability. Report by the International Institute for Applied Systems Analysis (IIASA), Vienna.

Fisher, A. C., Hanemann, W.M., Roberts M.J. and Schlenker, W. (2012). The Economic Impacts of Climate Change: Evidence from Agricultural Output and Random Fluctuations in Weather: Comment. *American Economic Review* 102: 3749–3760.

Foley, J.A., Ramankutty, N., Brauman, K.A., Cassidy, E.S., Gerber, J.S., Johnston, M., Mueller, N.D., O'Connell, C., Ray, D.K., West, P.C., Balzer, C., Bennett, E.M., Carpenter, S.R., Hill, J., Monfreda, C., Polasky, S., Rockström, J., Sheehan, J., Siebert, S., Tilman, D. and Zaks, D.P.M. (2011). Solutions for a Cultivated Planet. *Nature* 487: 337–342.

Fresco, L. and Poppe, K. (2016). Towards a common agricultural and food policy. Position paper on the future of CAP. Research report 09/2016, Wageningen University and Research. Available at: www.wur.eu/economic-research.

Fuchs, D., Kalfagianni, A. and Havinga, T. (2011). Actors in Private Food Governance: The Legitimacy of Retail Standards and Multistakeholder Initiatives with Civil Society Participation. *Agriculture and Human Values* 28(3): 353–367.

Galli, F. and Brunori, G. (2013). Short food supply chains as drivers of sustainable development. Evidence document, developed in the framework of the European Community FP7 project Foodlinks (GA No. 265287), Laboratorio di studi rurali Sismondi.

Geist, H., Lambin, E., McConnell, W.D.A. and Alves, D. (2005). Causes, Trajectories and Syndromes of Land-Use/Cover Change. *Newsletter of the International Human Dimensions Programme on Global Environmental Change* 3/2005.

Goodman, D. (2003). The Quality "Turn" and Alternative Food Practices: Reflections and Agenda. *Journal of Rural Studies* 19:1–7.

Gottlieb, R. (2001). *Environmentalism Unbound: Exploring New Pathways for Change.* Cambridge, MA: MIT Press.

Gottlieb, R. and Joshi, A. (2010). *Food Justice.* Cambridge, MA: MIT Press.

Grolleaud, M. (2002). Post-harvest losses: Discovering the full story. Overview of the phenomenon of losses during the post-harvest system. Agro-Industries and Post-Harvest Management Service, FAO, Rome, Italy.

Gunther, F. (2005). A solution to the heap problem: The doubly balanced agriculture: Integration with population. Available at: www.holon.se/folke/kurs/Distans/Ekofys/Recirk/Eng/balanced.shtml.

Hanjra, M.A. and Qureshi, M.E. (2010). Global Water Crisis and Future Food Security in an Era of Climate Change. *Food Policy* 35: 365–377.

Hassanein, N. (2003). Practicing Food Democracy: A Pragmatic Politics of Transformation. *Journal of Rural Studies* 19(1): 77–86.

Hassanein, N. (2008). Locating Food Democracy: Theoretical and Practical Ingredients. *Journal of Hunger & Environmental Nutrition* 3(2–3): 286–308.

Havinga, T., van Waarden, F. and Casey, D. (eds.) (2015). *The Changing Landscape of Food Governance: Public and Private Encounters.* Cheltenham, UK: Edward Elgar.

Hayes-Conroy, J. (2014). *Savoring Alternative Food: School Gardens, Healthy Eating and Visceral Difference.* London and New York: Routledge.

Heinrichs, C. (2003). The Practice and Politics of Food Systems Localization. *Journal of Rural Studies* (9): 33–45.

Hole, D.J., Perkins, A.J., Wilson, J.D., Alexander, I.H., Grice, P.D. and Evans, A.D. (2006). Does Organic Farming Benefit Biodiversity? *Biological Conservation* 122: 113–130.

Holloway, L., Kneafsey, M., Venn, L., Cox, R., Dowler, E. and Tuomainen, H. (2007). Possible Food Economies: A Methodological Framework for Exploring Food Production–Consumption Relationships. *Sociologia Ruralis* 47:1–19.

Hunt, A.R. (2015). *Civic Engagement in Food System Governance: A Comparative Perspective of American and British Local Food Movements*. London and New York: Earthscan.

Ilbery, B. and Maye, D. (2005). Alternative (Shorter) Food Supply Chains and Specialist Livestock Products in the Scottish English Borders. *Environment and Planning A* 37:823–844.

IPCC (2007). Climate change 2007: Synthesis report. Contributions of working groups I, II and III to the fourth assessment report of the Intergovernmental Panel on Climate Change. IPCC, Geneva.

Kalfagianni, A. (2014). Addressing the Global Sustainability Challenge: The Potential and Pitfalls of Private Governance from the Perspective of Human Capabilities. *Journal of Business Ethics* 122(2): 307–320.

Khan, S. and Hanjra, M.A. (2009). Footprints of Water and Energy Inputs in Food Production – Global Perspectives. *Food Policy* 34(2): 130–140.

Khan, S., Hanjra, M.A. and Mu, J. (2009a). Water Management and Crop Production for Food Security in China: A Review. *Agricultural Water Management* 96(3): 349–360.

Khan, S., Khan, M.A., Hanjra, M.A. and Mu, J. (2009b). Pathways to Reduce the Environmental Footprints of Water and Energy Inputs in Food Production. *Food Policy* 34(2): 141–149.

King, R.P., Hand, M.S., DiGiacomo, G., Clancy, K., Gomez, M.I., Hardesty, S.D., Lev, L. and McLaughlin, E.W. (2010). Comparing the structure, size and performance of local and mainstream food supply chains. Economic Research Report No. 99, USDA, Washington.

Kneafsey, M., Schmutz, U., Balasz, B. and Trenschard, L. (2013). Short food supply chains and local food systems in the EU. A state of play of their socio-economic characteristics. JRC Scientific and Policy Report 06/2013, European Commission, Luxembourg.

Lang, T. (1998). Towards a Food Democracy. In: S. Griffiths and J. Wallace (eds.), *Consuming Passions: Food in the Age of Anxiety*. Manchester: Mandolin, Manchester University Press, pp. 13–23.

Lang, T. and Heasman, M. (2004). *Food wars: The Global Battle for Mouths, Minds and Markets*. London: Earthscan.

Levkoe, C.Z. (2011). Towards a Transformative Food Politics. *Local Environment* 16(7): 687–705.

McMichael, P. and Schneider, M. (2011). Food Security Politics and the Millennium Development Goals. *Third World Quarterly* 32(1): 119–139.

Mariola, M.J. (2008). The Local Industrial Complex? Questioning the Link between Local Foods and Energy Use. *Agriculture and Human Values* 25(2): 193–196.

Marsden, T., Banks, J. and Bristow, G. (2002). The Social Management of Rural Nature: Understanding Agrarian-Based Rural Development. *Environment and Planning A* 34: 809–825.

Matson, P.A., Parton, W.J., Power, A.G. and Swift, M.J. (1997). Agricultural Intensification and Ecosystem Properties. *Science* 277: 504–509.

Molden, D. and Fraiture, C.D. (2004). Investing in water for food, ecosystems and livelihoods. Blue paper, Comprehensive Assessment of Water Management in Agriculture, Stockholm.

Molden, D., Oweis, T.Y., Steduto, P., Kijne, J.W., Hanjra, M.A., Bindraban, P.S., Bouman, B.A.M., Cook, S., Erenstein, O., Farahani, H., Hachum, A., Hoogeveen,

J., Mahoo, H., Nangia, V., Peden, D., Sikka, A., Silva, P., Turral, H., Upadhyaya, A. and Zwart, S. (2007). Pathways for Increasing Agricultural Water Productivity. In: D. Molden (ed.), *Water for Food, Water for Life: A Comprehensive Assessment of Water Management in Agriculture*. London: Earthscan and Colombo: International Water Management Institute.
Molden, D., Oweis, T., Steduto, P., Bindraban, P., Hanjra, M.A. and Kijne, J. (2010). Improving Agricultural Water Productivity: Between Optimism and Caution. *Agricultural Water Management* 97(4): 528–535.
Morgan, E. (2009). Fruit and vegetable consumption and waste in Australia. Victorian Health Promotion Foundation, State Government of Victoria, Australia.
Morgan, K. (2010). Local and Green, Global and Fair: The Ethical Foodscape and the Politics of Care. *Environment and Planning A* 42:1852–1867.
Morgan, K.J. (2015). Nourishing the City: The Rise of the Urban Food Question in the Global North. *Urban Studies* 52(8): 1379–1394.
Morgan, K. and Sonnino, R. (2010). The Urban Foodscape: World Cities and the New Food Equation. *Cambridge Journal of Regions, Economy and Society* 3(2): 209–224.
Morgan, K., Marsden, T. and Murdoch, J. (2006). *Worlds of Food: Place, Power and Provenance in the Food Chain*. Oxford: Oxford University Press.
NEF (2010). The great transition: A tale of how it turned out right. New Economics Foundation report. Available at: https://issuu.com/neweconomicsfoundation/docs/great_transition.
O'Hara, J.K. (2011). *Market Forces. Creating Jobs through Public Investment in Local and Regional Food Systems*. Cambridge, MA: Union of Concerned Scientists.
Oglethorpe, D. (2009). Investigating the practicalities and benefits of local food production, and identifying any unintended effects and trade-offs. Project report, Defra, London.
Parfitt, J., Barthel, M. and MacNaughton, S. (2010). Food Waste within Food Supply Chains: Quantification and Potential for Change to 2050. *Philosophical Transactions of the Royal Society B* 365: 3065–3081.
Pimbert, M. (2009). *Towards Food Sovereignty: Reclaiming Autonomous Food Systems*. London: International Institute for Environment and Development.
Pirog, R., Miller, C., Way, L., Hazekamp, C. and Kim, E. (2014). The local food movement: Setting the stage for good food. Center for Regional Food Systems, Michigan State University.
Plassmann, K. and Edwards-Jones, G. (2009). *Where Does the Carbon Footprint Fall? Developing a Carbon Map of Food Production*. London: International Institute for Environment and Development.
Pretty, J.N., Ball, A.S., Lang, T. and Morison, J.I.L. (2005). Farm Costs and Food Miles: An Assessment of the Full Cost of the UK Weekly Food Basket. *Food Policy* 30: 1–19.
Prugh, T., Costanca, R. and Daly, H. (2000). *The Local Politics of Global Sustainability*. Washington, DC: Island Press.
Rayfuse, R. and Weisfelt, N. (eds.) (2012). *The Challenge of Food Security: International Policy and Regulatory Frameworks*. Cheltenham, UK: Edward Elgar.
Reardon, T., Henson, S. and Berdegué, J. (2007). "Proactive Fast-tracking" Diffusion of Supermarkets in Developing Countries: Implications for Market Institutions and Trade. *Journal of Economic Geography* 7(4): 399–432.
Renting, H., Marsden, T. and Bank, J. (2003). Understanding Alternative Food Networks: Exploring the Role of Short Food Supply Chains in Rural Development. *Environment and Planning A* 35: 393–411.

Sage, C. (2002). Social Embeddedness and Relations of Regard: Alternative "Good Food" Networks in South-West Ireland. *Journal of Rural Studies* 19: 47–60.

Sage, C. (2013). The Interconnected Challenges for Food Security from a Food Regimes Perspective: Energy, Climate and Malconsumption. *Journal of Rural Studies* 29: 71–80.

Santini, F., Gomez, S. and Paloma, G. (eds.). (2013). Short food supply chains and local food systems in the EU. A state of play of their socio-economic characteristics. Joint Research Centre Scientific and Policy report EUR 25911 EN, European Commission, Luxembourg.

Sbicca, J. (2012). Growing Food Justice by Planting an Anti-Oppression Foundation: Opportunities and Obstacles for a Budding Social Movement. *Agriculture and Human Values* 29: 455–466.

Scott, A. (1990). *Ideology and the New Social Movements*. London: Unwin Hyman.

Seems, R. (2010). Putting Place on the Menu: The Negotiation of Locality in UK Food Tourism, from Production to Consumption. *Journal of Rural Studies* 26:105–115.

Seyfang, G. (2008). Avoiding Asda? Exploring Consumer Motivations in Local Organic Food Networks. *Local Environment* 13(3): 187–201.

Sharzer, G. (2012). *No Local: Why Small-Scale Alternatives Won't Change the World*. Alresford, UK: Zero Books.

Smil, V. (2000a). *Feeding the World: A Challenge for the 21st Century*. Cambridge, MA: MIT Press.

Smil, V. (2000b). Phosphorus in the Environment: Natural Flows and Human Interferences. *Annual Review of Energy and the Environment* 25: 53–88.

Sonnino, R. and Marsden, T. (2006). Beyond the Divide: Rethinking Relationships between Alternative and Conventional Food Networks in Europe. *Journal of Economic Geography* 6: 181–199.

Spaargaren, G., Oosterveer, P. and Loeber, A. (eds.) (2012). *Food Practices in Transition: Changing Food Consumption, Retail and Production in the Age of Reflexive Modernity*. London and New York: Routledge.

Thönissen, R. (2009). Food waste: The Netherlands. Presentation to the EU Presidency Climate Smart Food Conference, November 2009, Lund, Sweden.

Tregear, A. (2011). Progressing Knowledge in Alternative and Local Food Networks: Critical Reflections and a Research Agenda. *Journal of Rural Studies* 27(4): 419–430.

United Nations (2015). Transforming our world: The 2030 agenda for sustainable development. UN Resolution A/RES/70/1.

USEPA (United States Environmental Protection Agency) (2009). Municipal solid waste generation, recycling, and disposal in the United States: Facts and figures for 2008. Available at: www.paperbecause.com/PIOP/files/48/48417708-7e64-45a7-9720-d2c50e00f3be.pdf.

van der Ploeg, J.D. (2010). The Food Crisis, Industrialized Farming and the Imperial Regime. *Journal of Agrarian Change* 10(1): 98–106.

Van Loo, E.J., Caputo, V., Nayga Jr., R.M., Seo, H-S., Zhang, B. and Verbeke, W. (2015). Sustainability Labels on Coffee: Consumer Preferences, Willingness-to-Pay and Visual Attention to Attributes. *Ecological Economics* 118: 215–225.

Via Campesina (2003). What is food sovereignty? Available at: www.viacampesina.org/IMG/_article_PDF/article_216.pdf.

Watts, D., Ilbery, B. and Jones, G., (2007). Networking Practices among "Alternative" Food Producers in England's West Midlands Region. In: D. Maye, L. Holloway and M. Kneafsey (eds.), *Alternative Food Geographies*. Bingley, UK: Emerald Group Publishing.

Welsh, J. and MacRae, R. (1998). Food Citizenship and Community Food Security: Lessons from Toronto, Canada. *Canadian Journal of Development Studies* 19: 237–255.

WHO (2018). Obesity and overweight: Key facts. Available at: www.who.int/newsroom/fact-sheets/detail/obesity-and-overweight.

Wilkins, J.L. (2005). Eating Right Here: Moving from Consumer to Food Citizen. *Agriculture and Human Values* 22: 269–273.

Williams, A.G., Audsley, E. and Sandars, D.L. (2006). Determining the environmental burdens and resource use in the production of agricultural and horticultural commodities. Defra research project IS0205, Cranfield University and Defra.

Windfuhr, M. and Jonsén, J. (2005). Food Sovereignty: Towards democracy in localized food systems. The Schumacher Centre for Technology and Development, Rugby, UK.

Wiskerke, J. (2009). On Places Lost and Places Regained: Reflections on the Alternative Food Geography and Sustainable Regional Development. *International Planning Studies* 14(4): 369–387.

Wittman, H. (2011). Food Sovereignty: A New Rights Framework for Food and Nature? *Environment and Society: Advances in Research* 2(1): 87–105.

Wood, S., Ehui, S., Alder, J., Benin, S., Kassman, K.G., Cooper, H.D., Johns, T., Gaskell, J., Grainger, R., Kadungure, S., Otte, J., Rola, A., Watson, R., Wijkstrom, U. and Devendra, C. (2005). Chapter 8: Food. In: Millennium Ecosystem Assessment report, *Ecosystems and Human Well-being: Current State and Trends*. Washington, DC: Island Press.

World Commission on Environment and Development (1987). *Our Common Future*. Oxford: Oxford University Press.

Part I

Innovative local agri-food governance

Short food supply chains (SFSCs) are often perceived as new and innovative forms of local governance combining entrepreneurship with local values while serving the public good. Yet, how is innovation understood exactly? What are its different manifestations? And what are the consequences for food sustainability? The chapters in this section aim to provide answers to these questions. Ranging from municipal food movements formed around the concept of "good food", to grassroots initiatives and to entrepreneurial forms of food localization, innovation takes multiple forms and shapes and has multiple consequences. Importantly, as the chapters in this section demonstrate, these depend largely on context; on the ability of food initiatives to transcend their differences and form a common identity; on safeguarding themselves from efforts to co-opt them; on embracing ethnic and others forms of diversity; and on embracing sustainability, not only in its environmental but also its economic and social dimensions, particularly concerns about equity and justice.

2 The rise of municipal food movements

Kevin Morgan and Raychel Santo

Introduction: rendering food visible

If all the food-based movements across the globe were stripped of their single-issue status and assigned a common designation – such as the "Good Food Movement", for example – it might transpire to be the largest and fastest growing social movement in the world. Many of these movements seem to be dedicated to a single purpose – hunger relief, urban agriculture, diet-related disease prevention, economic regeneration, land access, fair trade, animal welfare or rights, pollinator protection, school food reform, food sovereignty, food worker rights, climate change mitigation, food tradition and culture preservation, tackling antibiotic resistance, and so forth (Morgan, 2015b). Indeed, until academic (Hunt, 2015; Marsden and Franklin, 2013) and popular (Latham, 2016; Pollan, 2010) writers began highlighting the rise of a heterogeneous food movement, we often failed to recognize that food is the common denominator of so much social and political activity around the world today. But there are deeper and more powerful forces preventing us from seeing the centrality of food in our lives, and perhaps the most important of these is the industrial food system itself. Here is a system that has many achievements to its credit, not least the awesome productivity gains that until recently allowed the industrial food system to claim that it had conquered hunger in the countries of the Global North. On the other side of the ledger, however, this is a system that has shattered the traditional bonds between product and place, and between production and consumption, as commodified food markets have become ever more globalized and foodscapes ever more placeless. Despite its many feats, the costs of the industrial food system have become more and more manifest in recent decades and the principal ones are to be found in the noxious cocktail of diet-related disease, ecological damage and social injustice.

The Good Food Movement(s) as a whole, which some refer to as the rise of alternative (Goodman *et al.*, 2012) or civic (Renting *et al.*, 2012) food networks, have been the most conspicuous response to these problems and – for all their differences – they have sought to *re-moralize* the food system by contesting the logic that aims to reduce food to the status of a pure commodity.

One of the great dreams of neoliberalism has been its concerted campaign to "normalize" the agri-food sector by claiming it is an industry like any other and therefore it should be treated in the same way in multilateral trade negotiations as we treat automobiles, steel and software. But contrary to the neoliberal mantra that food is just another business, food cannot be commodified so easily. While food is, of course, a business, it is a business like no other. Because we ingest the products of the agri-food sector, food plays a role in health and wellbeing – of people and planet alike – that differentiates it from any other manufactured product or service (Morgan, 2015a). Indeed it is no coincidence that the concept of *moral economy* was originally developed in the context of a study of crowds to explain why food had a unique capacity to trigger riots (Thompson, 1971).

Today the concept of moral economy has evolved into a powerful lens through which to convey the ethical rights and responsibilities that we have towards each other and to nature. It also allows us to critique standards and practices that jeopardize public health, ecological integrity and social justice, the intrinsically significant values at the heart of a capacious conception of sustainability (Morgan, 2015a; Sayer, 2007). As part of the larger Good Food Movement(s), municipal food movements are at the forefront of campaigns to re-moralize the food system by highlighting the unique attributes of the agri-food sector and the critical role that *good* food plays in sustaining the health and wellbeing of people and planet. To do so, municipal innovators – most notably in North America and the UK but increasingly across other parts of Europe and the world too – within local governments, civil society groups, academia and businesses have been overcoming fragmentation in "sector, level of activity, discipline, and narrowly-defined objectives" and exclusionary structures to build long-term systemic solutions in local food strategies, policies and governance practices (Hunt, 2015: 216). We use the term "municipal" instead of "urban" to emphasize that emerging innovations in local food governance are not confined to city-regions; municipalities in many rural and suburban areas are also revolutionising traditional food governance regimes.

These municipal food movements are analysed in a more granular fashion in the following section. The remainder of this opening section highlights three questions that loom large in the debate about the nature of the Good Food Movement(s) as a whole, and we return to these questions in the final section when we assess the progressive credentials of these social movements.

The scalar question

Notwithstanding international food movements such as Via Campesina, the vast majority of food-focused social movements are largely *sub-national* phenomena, based as they are on jurisdictions that range from small towns at one end of the spatial spectrum to cities and states at the other end. Although the "local" label can often conceal as much as it reveals, the key point to establish is that these sub-national spaces of collective action have been mobilized to fill the vacuum left by the inaction of national and international agri-food institutions and players (Moragues-Faus and Morgan, 2015: 1566; Sheingate, 2014). Far from being bounded spaces, they are highly porous multi-scalar formations because they are interpolated with and allied to extra-local currents and movements. Municipal food movements have been one of the most prominent examples of local food movements, partly because of burgeoning urbanisation and partly because national political landscapes are increasingly municipality-centric. The scalar challenge for a municipal food movement is how to become more than a localist voice in national and international food politics, an issue we address in the final section.

The sectoral question

The central focus of Good Food Movement(s) seems to have shifted over time from an early preoccupation with production-related issues (e.g. sustainable agriculture, organic certification, etc.) to a much more explicit focus on consumption-led activities (e.g. healthy eating and food access/food justice issues). This shift in focus mirrors the earlier shift in international food policy discourse triggered by the seminal work of Amartya Sen on poverty and famines which demonstrated that famine is a condition where people do not have enough food to eat, rather than there not being enough food. Thus, access to (rather than supply of) food was the critical issue. Far from being a purely academic matter, the shift towards a food access lens could save many lives because the traditional "food availability approach" has "deranged anti-famine policies in the past" (Sen, 1982: 8). Consumption-focused food movements also have a much wider resonance because consumers dwarf producers numerically and in political significance. The sectoral challenge for Good Food Movement(s) is how to transcend binary divisions between consumption and production and to recalibrate the linkages in a more sustainable form (e.g. institutional procurement policies, short supply chains, farmers markets, food hubs, box schemes, etc.) (Renting *et al.*, 2012).

The social question

Good Food Movement(s) as a whole have been severely criticized for being blind to class, race and gender as well as for green parochialism (Alkon and Agyeman, 2011; Guthman, 2008; Morgan, 2013). The social question raises profoundly important questions about the social composition of food movements as well as the kind of issues that they campaign around. For example, the problem of undignified wages and rights among farmworkers, food service and other food system workers has been largely ignored until very recently – indeed, until they began organising to expose the issues through initiatives such as the Food Chain Workers Alliance in the US (FCWA, 2012). The fact that the vast majority of fast food workers earn around the minimum wage and nearly 90 per cent have no health benefits, driving them to rely on publicly-funded safety-net programmes to make ends meet, has fuelled the US campaign for $15/hour (Allegretto *et al.*, 2013). The campaign has gained traction among mainstream politicians from Clinton to Cuomo, providing a good illustration of how local campaigns can morph into national campaigns. That said, it is worth noting that local food movements do not need to wait for national action to validate their claims because, in the case of low pay, the cities of Seattle and San Francisco and the states of New York and California have already voted for $15/hour. The low pay issue also underlines the critically important interplay between civil society organisations on the one hand and local government institutions on the other because this combination – of municipal activism and civic engagement – can help local food movements become more than a protest voice. The remainder of this chapter is organized as follows: the second section explores the growth of municipal food movements in the UK and the US, highlighting the nuances of scale and scope, while the concluding section assesses the implications of the analysis, especially with respect to the progressive credentials of municipal food movements.

UK and US municipal food movements

Rise of municipal food policy groups

Epitomising the various components and actors of municipal food movements are local, place-based initiatives in food governance and policy – notably food (policy) councils, partnerships, coalitions and other similar groups, hereafter called local food policy groups (LFPGs) as coined by Halliday (2015). Since the first LFPG began in Knoxville, Tennessee in the mid-1980s, followed by other key pioneers in London and Toronto, LFPGs have been emerging across the Global North, most rapidly over the last decade (see Figure 2.1). They are most heavily concentrated in the US (284), UK (55) and Canada (52) (CLF, 2018; SFCN, 2018), though some also exist or are developing in Germany, Belgium, New Zealand and several other European countries (Balázs, 2012; Forster *et al.*, 2015; Heuser *et al.*, 2015). These groups assemble food system stakeholders from government (e.g. city council, administrators), civil society (e.g. NGOs, local residents) and the private sector (e.g. local businesses, farmers) to evaluate, influence and – in some cases – create food policy and programmes, as well as foster new relationships and interconnections between food system initiatives at the municipal and (in North America) state/province, regional and First Nations or Native American levels.[1] Changes in public food procurement procedures, land use planning codes to support (peri-)urban agriculture, improved access to healthier food, reduced and redirected food waste, support for the development of food hubs and a vibrant local food economy and increased public education are just some of the improvements that have resulted from this increased civic engagement in food systems (Carey, 2015; Essex *et al.*, 2015).

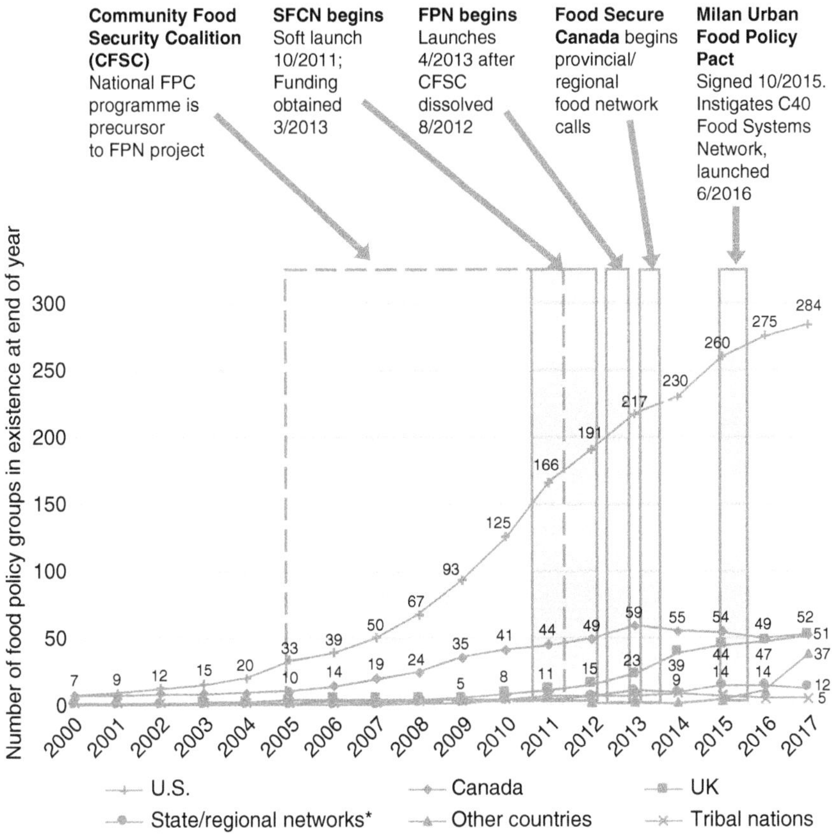

Figure 2.1 Rise of LFPGs and the networks which connect them globally.

Source: created by author using data compiled from CLF's annual FPC directory update, SFCN website, websites and Facebook pages for LFPGs and state/regional networks, and personal communication. Other countries developing LFPGs (as defined on p. 30) include Germany, Belgium, New Zealand, Spain, Austria, France, Switzerland and the Netherlands.

Notes

1 State/regional networks were included if they play a role in convening, training and/or instigating LFGPs (>3) beyond any role focused on influencing state-level policy.

2 Thirteen LFGPs in the US, and 24 in "other countries" included in the 2017 figures, self-reported as being still in development. Thirty-one LFPGs in the US and 10 in Canada also reported as being in transition (redefining the purpose and/or structure of their group).

LFPGs in North America have mostly been independently instigated and supported (Harper *et al.*, 2009; Winne, 2013). Responding to calls for more collaboration between isolated municipal progress (Clancy, 2014), Johns Hopkins Center for a Livable Future launched its Food Policy Networks (FPN) project in 2013[2] to connect the efforts and build the capacity of existing LFPGs through providing (mostly online) networking platforms, a consolidated database of food policy resources, webinars and trainings on common thematic (e.g. procurement) and organizational (e.g. funding) issues, targeted action research and other technical assistance. Meanwhile, their British counterparts have been spurred by national leadership through the Sustainable Food Cities Network (SFCN), established in 2012 by a coalition of three NGOs – the Soil Association, Sustain and Food Matters – to help "people and places share challenges, explore practical solutions, and develop best practices" by offering frameworks for food strategies and action plans, a national networking conference, thematic webinars and a progress-based award scheme (SFCN, 2016).

SFCN and FPN are not the only trans-local networks of municipal food system reformers in the UK or North America. Some LFPGs engage with global networks (e.g. Milan Food Policy Pact, FAO Food-for-Cities listserv) and/or more regional ones (e.g. Sustain Ontario, Chesapeake Food Policy Leadership Initiative, EAT Nordic Cities Initiative, and 12 state[3] networks of LFPGs in the US). Other organisations also support municipal food system change. In the US, these include: Growing Food Connections project, Extension Communities of Practice network, Harvard Food Law and Policy Clinic, Center for Good Food Purchasing, American Planning Association Food Systems Interest Group, Conference of Mayors Food Policy Task Force, ChangeLab Solutions, Sustainable Economies Law Center and University of Arkansas Law School's Indigenous Food and Agriculture Initiative. In Canada, these include Food Secure Canada's provincial/territorial food security networks teleconference calls;[4] and in Europe, the TRANSMANGO research project, EUROCITIES Working Group on Urban Food Policy, CITYFOOD Initiative and EAT Forum.

Moreover, LFPGs are not the only mechanisms through which municipalities are effecting changes in local food system planning, policies and programmes, especially outside of North America and the UK. Many cities have embedded innovative food system reform into their government policies and agencies without prompting from or the subsequent creation of civil society engaged LFPGs, including Bangkok, Belo Horizonte, Malmö, Melbourne, Milan, Montpellier, Tukums, Turin and Windhoek (Deakin *et al.*, 2016; Forster *et al.*, 2015; Moragues-Faus and Morgan, 2015; Moragues-Faus *et al.*, 2013). While they may not have created long-term participatory food system governance spaces in the form of LFPGs, many of these cities have consulted with the public and/or expert stakeholders in, or following, the development of food strategies. The global development of LFPGs will likely continue with the inception of the Milan Urban Food Policy Pact (MUFPP), launched in October 2015 with 45 original signatories and endorsed by 167 cities as of June 2018. Mayors and leaders that sign the MUFPP affirm their commitment to advancing more *sustainable diets* – that is to say, diets that are healthy, safe, culturally appropriate, environmentally friendly and rights-based – within their cities. From the rapid growth in interest since its inception, it is evident that this is an issue that resonates worldwide. The Pact builds upon the successes of LFPGs, and encourages their expansion into additional cities, through one of its key recommendations to "enhance stakeholder participation at the city level through political dialogue, and if appropriate, appointment of a food policy advisor and/or development of a multi-stakeholder platform or food council, as well as through education and awareness raising" (MUFPP, 2018). The nascent C40 Food Systems Network was launched in June 2016 to support MUFPP commitment implementation through knowledge sharing and capacity building among members (Watts, 2016). The extent to which these global networks of LFPGs and other municipal food system reformers interact with already existing networks such as SFCN and FPN is still limited; more extensive future collaboration remains to be seen.

Multi-scalar, porous reach

LFPGs in the US and UK engage in work and relationships outside of single spatial scales, although they are commonly conceived of (e.g. names, jurisdictional boundaries) and analyzed within such confines. By working across multiple scales simultaneously, LFPGs may thus be considered one mechanism through which urban–rural and local–global divides are being blurred. For instance, they have been commended for avoiding parochialism through the incorporation of regional foodsheds and fair international trade flows as part of their food strategies (Mendes and Sonnino, 2018). While many are concentrated within urban centres, LFPGs may often include food producing representatives from outside city limits as members, and these representatives may be helping to shape the narratives and goals of the LFPGs beyond a defensive localism. In addition, more US LFPGs are jurisdictionally affiliated within counties (36 per cent) and not just cities (20 per cent) (CLF, 2018), despite their typical categorization as part of an "urban food movement" (e.g. Goodman *et al.*, 2012: 19; Morgan, 2015a). Many (15 per cent) even officially straddle the line between both city and county jurisdictions (CLF, 2018). In the UK, SFCN has begun to drive collective policy advocacy at the national level through its campaigns addressing food poverty and procurement policies. Meanwhile, the FPN project has initiated less structured campaigns for national level advocacy around specific issues and concentrated more on developing LFPGs' capacity to organize collective campaigns themselves. Questions remain over the extent to which their diverse structures, member compositions, funding sources, framings and activities constrain the ability of LFPGs to synthesize an identity strong enough to successfully advance collective action at higher levels (Santo and Moragues-Faus, in press).

Systems talk but systems take?

LFPGs in both the US and UK position themselves as cross-sectoral partnerships, working to address a wide range of food system issues from a variety of angles including health, sustainability, social justice and economic development.

Most groups in the US that have funding of some sort, however, receive it from public health-focused foundations or agencies (local, state and federal) (CLF, 2018). Likewise, much of the funding for UK local food system projects has come from public health-affiliated agencies (Hunt, 2015), with the exception of SFCN, which has received funding from a family foundation. While these funding realities may demonstrate the flexibility of LFPGs to adapt to current political and funding climates (which have been steeped in anti-obesity narratives), they could also be narrowing the scope of the work that LFPGs address at the expense of more holistic lenses.

Different priorities for diversity

In response to critiques of the Good Food Movement for re-producing structural inequities through its predominantly white, middle-class membership (Alkon and Agyeman, 2011; Guthman, 2008; Slocum and Cadieux, 2015), achieving diverse and meaningful community representation has become a common priority on which LFPGs (Day Farnsworth, 2017; McCullagh and Santo, 2014), and their funders (Hunt, 2015:190), concentrate. The extent to which this is being achieved is still up for debate, but it is an agenda item for nearly every group in the US. In contrast, UK food movement culture does not emphasize the empowerment/inclusion of people of diverse races, classes, genders and ages as much as it emphasizes "diverse organizational or professional representation" (Halliday, 2015: 206). Different membership compositions inherently lead to different policy and programmatic priorities of the LFPGs. For instance, in the US, some groups focused on incorporating and empowering marginalized community members have deliberately changed their names from "food policy councils" to more engaging names and aims. Meanwhile, in the UK, rather than directly organizing and engaging those who are experiencing food poverty, work addressing issues that impact lower-income communities and communities of colour tends to be led by professional organizations and networks that could influence policies and programmes related to living wages and increased access to healthy, affordable food.

Conclusions: key issues for food movement politics

In this final section we draw on the foregoing analysis to address four issues that continue to haunt food-based social movements, namely: (i) the definitional issue as to whether these social movements can be said to constitute a generic "Good Food Movement" when they exhibit such diversity; (ii) the challenge of co-optation by political elites and powerful lobbies such as public health; (iii) the spectre of green parochialism, which highlights the tensions within the "ethical foodscape" when sustainability narratives crowd out social justice narratives; and (iv) how to ensure that locally-embedded municipal food movements can engage with and/or be part of national and international food policy networks.

Food movement or food movements?

We began this chapter with a thought experiment in which we suggested that the food movement would probably constitute the largest and fastest growing social movement in the world if all single-issue food groups operated under a common designation like the "Good Food Movement". Although the definition of food quality is a highly contested and context-dependent question, the definition of "good food" in this chapter follows the simple but capacious definition employed by the Scottish Government in its *Good Food Nation* strategy, namely "food which is both tasty to eat and nutritious, fresh and environmentally sustainable"[5] (Scottish Government, 2014). This prosaic definition aims to convey the message that "good food" is aesthetically appealing, culturally appropriate and healthy for people and planet alike – attributes that would seem to span many of the concerns on the food issue spectrum. If this is the case, then it is not inconceivable that singe-issue food movements could subscribe to a more capacious cause, such as the cause of *Good Food for All*, which would be a complement to, rather than a substitute for, their own single–issue causes. In other words, a concerted campaign to build bridges between diverse food-based social movements would be part of a pluralist politics in which multiple identities are fashioned as part of a *cosmopolitan localism* (Morgan *et al.*, 2006). Fashioning multiple identities around food politics is to treat diversity in a positive sense – to see it as an asset to be nurtured rather than a liability to be jettisoned, which is how diversity is portrayed by the New Right in Trump's America and Brexit Britain.

If single-issue food movements are unable or unwilling to forge a common cause – if, in other words, they cannot fashion a more coherent Food Movement identity – then diversity could easily morph into fragmentation and fractiousness, the classic hallmarks of political failure. With their structural cross-sectoral diversity, LFPGs inherently offer an opportunity to assemble participants from multiple food causes into one space to collaborate and form a more coherent movement. How effectively they balance this nominal diversity, however, is still up for question, as groups report challenges in negotiating the contentious dynamics between members with different priorities (e.g. restaurants vs labour unions, conventional or industrial vs sustainable or small-scale farmers, hunger alleviation vs food access).

The challenge of co-optation

One of the putative signs of success is that many LFPGs are now being courted by local political elites because the latter are beginning to recognize that food policy issues resonate deeply with consumers and citizens, so much so that food policy provides a platform to help politicians reconnect with their publics. This creates a major dilemma for LFPGs – what we call the challenge of co-optation – because being involved in food policy-making might suggest that these social movements have finally secured some political influence, when in reality it might merely be the semblance of power not the substance. The worst-case scenario is where the politics of co-governance (e.g. LFPGs embedded within or working closely with local state bodies) degenerate into the politics of co-optation, where civil society groups "sacrifice their radical voice for the semblance of political influence" (Morgan, 2015a: 11).

But the challenge of co-optation can also come from other quarters, such as the powerful public health lobby that often sponsors local food campaigns as part of its programme to combat diet-related diseases such as obesity. Although it might be more benign than local political elites, the public health lobby can paradoxically neuter the political dynamism of food movements by reducing the food policy discourse to a narrow nutritional agenda, thereby transforming the food narrative from a sustainability narrative to a medical one. The scarcity of alternative funding options means that LFPGs will often be dependent on the public health community; therefore, the former need to assert their independence from the latter if they want to ensure that the values of social justice and ecological integrity are given parity of esteem with public health.

Green parochialism

As we suggested in the introductory section, one of the deepest and oldest tensions within food-based social movements is between two equally important causes – the cause of *sustainability* (where this is largely framed in environmental terms) and the cause of *social justice* (where this is framed in terms of advancing race, class and gender equity). This tension became so acute in the US that it was one of the main reasons why the most important food policy alliance, the Community Food Security Coalition, decided to disband in 2012. Given its origins in environmental campaigns, green activists have freely conceded that the organic food movement in the US was "largely blind to hunger, race and class issues" (Winne, 2008: 133). This tension also spawned a war of words between green campaigners and the international development community in the UK when the former invoked the (spurious) issue of food miles to promote local produce over imported produce from developing countries, even though transportation was a relatively small component of the carbon footprints of such products. This was a prime example of "green parochialism" where sustainability was framed entirely in environmental terms to the exclusion of the social justice claims of poor primary producers in developing countries (Morgan, 2010). This conflict within the "ethical foodscape" demonstrated that sustainability is a multidimensional cause, underlining the need for strategies that are sufficiently capacious and pluralist that they accommodate variety and foster unity-in-diversity, the holy grail of the food movement. As discussed in the second section, issues related to social inclusion and justice are still manifesting in debates over the composition and priorities of LFPGs, and will need to be adequately addressed if the undesired outcomes of "green parochialism" are to be avoided.

Beyond localism

By their very nature LFPGs are locally-focused social movements. This is both a strength and a weakness: a strength because it means they are embedded in their communities; a weakness because an exclusively local focus renders them vulnerable to national and international forces that may compromise their goals. The Marxist theorist David Harvey framed the dilemma well when he said that

> the contemporary emphasis on the local, while it enhances certain kinds of sensitivities, totally erases others and thereby truncates rather than emancipates the field of political engagement and action. While we all may have some "place" (or "places") in the order of things, we can never ever be purely "local" beings, no matter how hard we try.
>
> (Harvey, 1996: 195)

While this provides a stunningly concise critique of pure localism, it offers little by way of a prospectus as to how, in practical terms, social movements are supposed to combine local and non-local action in the multi-scalar systems that govern us today. Our analysis, however, suggests that food movements are at the forefront of campaigns that are intrinsically multi-scalar in the sense that they are neither purely local nor purely global. With the assistance of regional and national networks ranging from the SFCN in the UK and FPN in the US to the MUFPP at the global scale, LFPGs can remain locally embedded and also globally engaged. In doing so, municipal food movements can also help to redress two of the big weaknesses of the recently agreed Sustainable Development Goals, namely, the woeful absence of a sub-national *spatial sensibility* and the pitifully low status accorded to *implementation mechanisms*. Where they harness the concerted power of civic engagement and municipal activism, municipal food movements can bring their spatially-sensitive local knowledge to bear on the challenge of sustainable diets. They can also be one of the implementation agents to ensure that the grandiloquent rhetoric of the international development community (UN agencies, donors, etc.) actually gets translated into the prosaic things that satisfy human needs and enhance human capabilities.

Notes

1 This chapter concentrates most on LFPGs – which make up the majority (71 per cent) of LFPGs in North America (CLF, 2018) – though state/provincial (8 per cent), regional (20 per cent) and First Nations or Native American (1 per cent) food policy groups also exist.

2 FPN formed after the dissolution of the Community Food Security Network (CFSC) in 2012, in order to maintain and continue the local and state food policy work (including FPC-related) initiated by the CFSC. While CFSC served as a precursor to FPN, its services were meagre compared to FPN's resources and staff time.

3 This number only includes state networks which play a role in convening, training and/or instigating LFPGs (>3) beyond any role focused on influencing state-level policy.
4 Although it surveys Canadian LFPGs (a relic of CFSC's North American scope), FPN's efforts focus primarily on the US and hence this chapter concentrates on its role there. Food Secure Canada (beginning in 2013) also hosts quarterly teleconferences with representatives of provincial/territorial food security networks and organisations, though these services are limited compared to FPN's and SFCN's.
5 Note that this definition excludes discussion of support for food produced under ethical labour conditions, a core tenet of many food movement activists.

References

Alkon, A.H. and Agyeman, J. (2011). *Cultivating Food Justice: Race, Class, and Sustainability*. Cambridge, MA: MIT Press.

Allegretto, S., Doussard, M., Graham-Squire, D., Jacobs, K., Thompson, D. and Thompson, J. (2013). Fast food, poverty wages: The public cost of low-wage jobs in the fast food industry. Center for Labor Research and Education, University of California and Department of Urban & Regional Planning, University of Illinois. Available at: http://laborcenter.berkeley.edu/pdf/2013/fast_food_poverty_wages.pdf (accessed 24 September 2016).

Balázs, B. (2012). Local Food System Development in Hungary. *International Journal of Sociology of Agriculture and Food* 19(3), pp. 403–421.

Carey, J. (2015). The Bristol method: How to become a more sustainable food city. Bristol 2015 European Green Capital website. Available at: www.bristol2015.co.uk/method/food-nature/ (accessed 3 March 2016).

Clancy, K. (2014). Food System Governance. *Journal of Agriculture, Food Systems, and Community Development* 4(2), pp. 3–6.

CLF (Johns Hopkins Center for a Livable Future) (2018). Food policy council directory [Data file and personal communication, received 14 June 2018].

Day Farnsworth, L. (2017). Beyond Policy: Race, Class, Leadership, and Agenda-Setting on North American Food Policy Councils. In: C. Etmanski (ed.), *Food Leadership: Leadership and Adult Learning for Global Food Systems Transformation*. Rotterdam: Sense Publishers.

Deakin, M., Diamantini, D. and Borrelli, N. (eds.) (2016). *The Governance of City Food Systems: Case Studies from Around the World*. Milan: Fondazione Giangiacomo Feltrinelli.

Essex, A., Shinkle, D. and Bridges, M. (2015). *Harvesting Healthier Options: State Legislative Trends in Local Foods 2012–2014*. Denver, CO: National Conference of State Legislatures.

FCWA (Food Chain Workers Alliance) (2012). The hands that feed us: Challenges and opportunities for workers along the food chain. Available at: http://foodchainworkers.org/wp-content/uploads/2012/06/Hands-That-Feed-Us-Report.pdf (accessed 20 August 2018).

Forster, T., Egal, F., Renting, H., Dubbeling, M. and Escudero, A.G. (eds.) (2015). *Milan Urban Food Policy Pact: Selected Good Practices from Cities*. Milan: Fondazione Giangiacomo Feltrinelli. Available at: www.ruaf.org/sites/default/files/MUFPP_SelectedGoodPracticesfromCities.pdf (accessed 3 June 2016).

Goodman, D., DuPuis, E.M. and Goodman, M.K. (2012). *Alternative Food Networks: Knowledge, Practice, and Politics*. Abingdon, UK: Routledge.

Guthman, J. (2008). "If They Only Knew": Color Blindness and Universalism in California Alternative Food Institutions. *The Professional Geographer* 60(3), pp. 387–397.

Halliday, J.J. (2015). *A New Institutionalist Analysis of Local Level Food Policy in England between 2012 and 2014.* Doctoral dissertation, City University, London.

Harper, A., Shattuck, A., Holt-Giménez, E., Alkon, A. and Lambrick, F. (2009). *Food Policy Councils: Lessons Learned.* Oakland, CA: Food First Institute for Food and Development Policy.

Harvey, D. (1996). *Justice, Nature and the Geography of Difference.* Oxford: Blackwell.

Heuser, A., Pohl, C., Urhahn, J. and Buron, S. (2015). *Designing our Own Food: A Handbook on Food Policy Councils.* Berlin: INKOTA-netzwerk e.V [English translation received from C. Pohl by email on 16 August 2016].

Hunt, A.R. (2015). *Civic Engagement in Food System Governance: A Comparative Perspective of American and British Local Food Movements.* Oxford: Routledge.

Latham, J. (2016). Why the food movement is unstoppable. *Independent Science News.* Available at: www.independentsciencenews.org/health/why-the-food-movement-is-unstoppable (accessed 22 September 2016).

McCullagh, M. and Santo, R. (eds.) (2014). *Food Policy for All: Inclusion of Diverse Community Residents on Food Policy Councils.* Baltimore, MD: Center for a Livable Future.

Marsden, T. and Franklin, A. (2013). Replacing Neoliberalism: Theoretical Implications of the Rise of Local Food Movements. *Local Environment* 18(5), pp. 636–641.

Mendes, W. and Sonnino, R. (2018). Urban Food Governance in the Global North. In: T. Marsden (ed.). *Handbook of Nature.* London: Sage.

Moragues-Faus, A. and Morgan, K. (2015). Reframing the Foodscape: The Emergent World of Urban Food Policy. *Environment and Planning A* 47(7), pp. 1558–1573.

Moragues-Faus, A., Morgan, K., Moschitz, H., Neimane, I., Nilsson, H., Pinto, M., Rohracher, H., Ruiz, R., Thuswald, M., Tisenkopfs, T. and Halliday, J. (2013). Urban food strategies: The rough guide to sustainable food systems. Document developed in the framework of the European Commission FP7 project Foodlinks (GA No. 265287).

Morgan, K. (2010). Local and Green, Global and Fair: The Ethical Foodscape and the Politics of Care. *Environment and Planning A* 42(8), pp. 1852–1867.

Morgan, K. (2013). The Rise of Urban Food Planning. *International Planning Studies* 18(1), pp. 1–4.

Morgan, K. (2015a). Nourishing the City: The Rise of the Urban Food Question in the Global North. *Urban Studies* 52(8), pp. 1379–1394.

Morgan, K. (2015b). *Good Food for All.* Senedd Paper 3, Institute of Welsh Affairs, Cardiff, Wales.

Morgan, K.J., Marsden, T.K. and Murdoch, J. (2006). *Worlds of Food: Place, Power and Provenance in the Food Chain.* Oxford: Oxford University Press.

MUFPP (Milan Urban Food Policy Pact) (2018). *Recommended Actions.* Available at: www.milanurbanfoodpolicypact.org/mufpp_governance/ (accessed 7 September 2018).

Pollan, M. (2010). The food movement, rising. *New York Review of Books.* Available at: www.nybooks.com/articles/2010/06/10/food-movement-rising/ (accessed 14 September 2016).

Renting, H., Schermer, M. and Rossi, A. (2012). Building Food Democracy: Exploring Civic Food Networks and Newly Emerging Forms of Food Citizenship. *International Journal of Sociology of Agriculture & Food* 19(3), pp. 289–307.

Santo, R. and Moragues-Faus, A. (In press). Towards a Trans-Local Food Governance: Exploring the Transformative Capacity of Food Policy Assemblages in the US and UK. *Geoforum*.

Sayer, A. (2007). Moral Economy as Critique. *New Political Economy* 12(2), pp. 261–270.

Scottish Government (2014). *Becoming a Good Food Nation*. Scottish Government, Edinburgh.

Sen, A. (1982). *Poverty and Famines: An Essay on Entitlement and Deprivation*. Oxford: Clarendon Press.

SFCN (Sustainable Food Cities Network) (2016). *About*. Available at: http://sustainablefoodcities.org/ (accessed 15 May 2016).

SFCN (Sustainable Food Cities Network) (2018). *Find a City*. Available at: http://sustainablefoodcities.org/findacity (accessed 4 September 2018).

Sheingate, A. (2014). The new politics of food in the United States. Paper presented to ECPR General Conference, The New Politics of Food and Agriculture, Panel P059: De-Compartmentalising Agricultural Policy Making? Glasgow, 3–6 September 2014.

Slocum, R. and Cadieux, K.V. (2015). Notes on the Practice of Food Justice in the US: Understanding and Confronting Trauma and Inequity. *Journal of Political Ecology* 22, pp. 27–52.

Thompson, E.P. (1971). The Moral Economy of the English Crowd in the Eighteenth Century. *Past & Present* 50, pp. 76–136.

Watts, M. (2016). Mayors are changing the way we think about food. *Huffington Post*, 10 June. Available at: www.huffingtonpost.com/mark-watts-2/mayors-are-changing-the-way-we-think-about-food_b_10366716.html (accessed 14 June 2016).

Winne, M. (2008). *Closing the Food Gap: Resetting the Table in the Land of Plenty*. Boston, MA: Beacon Press.

Winne, M. (2013). UK keen on food policy councils [Mark Winne's food policy blog]. Available at: www.markwinne.com/uk-keen-on-food-policy-councils/ (accessed 5 March 2016).

3 Grassroots responsible innovation initiatives in short food supply chains

Cristina Grasseni

Introduction

Increasing numbers of citizens and consumers look at non-intensive agricultural practices as alternatives to intensive farming and its impact on climate change. They specifically rethink the logistics of food provisioning to provide localized and smaller scale alternatives.

The impact of the globalized food system on climate is widely known. In 2008, a Greenpeace report identified western styles of food provisioning, and specifically intensive farming for global distribution, as one of the main issues that needed to be tackled in order to temper climate change and health hazards. In particular, agriculture-related greenhouse gas emissions are a major contributor to global warming, including both indirect emissions arising from agricultural practices (taking into consideration production, transport and use of agricultural fertilizers) and direct emissions from specific agricultural practices (for example, livestock and manure management). Added to this, the expansion of agriculture into uncultivated areas increases emissions from deforestation and land conversion (Bellarby *et al.* 2008: 1–26). Recently the debate has focussed on the effects of increased meat demand and of global food logistics. While the latter extends greenhouse gas emissions beyond the mere scope of agricultural production because of the impact of travel and transportation of goods and foods, the former necessitates more land to produce animal feed, and fossil fuels to produce and transport feed (Bellarby *et al.*, 2008: 26–28).

Changing consumers' diet is only one of the ways in which provisioning styles are being rethought. While consuming less meat, for example, is a choice that can be pursued individually, shunning the commodity distribution chain that profits from and thus finances intensive farming as a global system requires more complex thinking and collective action. Styles of food provisioning are thus particularly relevant to the issue of sustainability. Short food supply chains can be considered an innovation challenge for contemporary societies – a challenge requiring counter-epistemologies and re-training, including peer-to-peer lifelong re-education. Many new "worlds of food" have emerged and coexist (Morgan *et al.* 2006): some forms of community supported agriculture (CSA) involve a palette of new skills, from permaculture and composting (Abrahamson and Bertoni 2014) to gathering, exchanging and cooking local and seasonal food (Faigel and Sinfield 2017). Direct-to-customer organic farming is being experimented with according to various formulas, ranging from farmers' markets to the French AMAPs (*Associations pour le Maintien d'une Agriculture Paysanne*). The Slow Food movement has inspired "convivia" and "slow cities" (*Cittaslow*). Various forms of animal-sharing and city farms have led to digital platforms and box delivery schemes.

Short food supply chains are thus reviewed in this chapter as responses to the agri-food system challenges that the book's editors take as their starting point, adding a critical overview of the idea of "responsible innovation" drawing on multidisciplinary literature (Hankins 2019). I propose that collective food procurement can be analysed as a form of grassroots innovation that enables consumers to take collective responsibility in the global food system. I illustrate this with two examples, considering their limits and potential.

Based on ethnographic research, my contribution has a regional focus. After a brief review of fieldwork observation of the local food movement in Massachusetts, USA (2012–2014), I focus on an Italian case study (2009–2011), GAS – *gruppi di acquisto solidale* or solidarity purchase groups – in the context of a diverse European local food movement. I view these as two examples of grassroots responsible innovation through short food supply chains. In particular, I highlight GAS's relational and epistemological potential with respect to forging new partnerships and fostering knowledge circulation, and also compare them with similar phenomena studied by others (notably the Slow Cities movement studied by Sarah Pink, 2009). I conclude by discussing the comparative aspects that should be developed in a longer study and look at the pros and cons of grassroots groups as drivers of responsible innovation in food chains.

A crisis of responsibility

Sustainable food is the subject of numerous academic conferences, publications and academic programmes spanning engineering, urban planning, design and business schools. In fact, "the problem with sustainability is that as it becomes a universally accepted value, its meaning is becoming increasingly nebulous" (Ubois 2011: 4). In other words, as Krause and Sharma (2012) demonstrate ethnographically in the case of Amherst in Massachusetts, sustainability can be co-opted by various social groups with opposing interpretations and intentions to argue in favour of measures in the service "of the common good" as *each* sees it. "Instead of being a force for uniting disparate interests in the stewardship of a common heritage, sustainability appears to have become a lightning rod for old antagonisms, deepening the rifts that have existed for quite a while" (Krause and Sharma 2012: 640). Sustainability is thus an increasingly contested terrain. The very meaning of the term ranges from environmental stewardship to the search for technological fixes to the multiple challenges of climate change, demographic pressures and the inefficient and wasteful provisioning of energy and natural resources. However, such interpretations of sustainability are limited and are often contested. A "triple bottom line" (Henriques and Richardson 2004) approach claims that sustainability does not just have an environmental meaning but an economic one (an enterprise must not only avoid pollution but must make a profit) as well as a social meaning: our livelihoods and models of development must not only be economically viable and environmentally friendly but also socially equitable. The idea of food sovereignty adds a political dimension to this.

A variety of citizens' initiatives have expressed a need for sovereignty over matters of sustainability, and of food in particular. They not only request that their food is sustainable (as determined by others); they wish to take responsibility for it. Alternative agri-food networks, diverse models of community food economies, citizens' mobilization against agro-criminal networks and models for transition are among these initiatives. They provide opportunities to contextualize grassroots understandings of innovation, sustainability and responsibility on the basis of an in-depth comprehension of their cultural diversity in different historical and geographical contexts, which highlight cultural, class and gender differences as well as, sometimes, political and ethnic marginalization.

Sustainable innovation in food production – as applied, for instance, to genetically modified (GM) crops, animal cloning or land fertility crises – has so far been discussed using the conceptual framework of risk "assessment" and "management" of technological innovation. Finding universal normative criteria for responsibility in food innovation is, however, difficult if not impossible, as this implies a universal value framework that can inspire political action (politics, at its best, being a participatory deliberation of the goals of the *polis*, namely, of the community of citizens). Social studies of science and technology provide a critical method for analysing the assumptions in which debates – or the absence of debate – are cast (Jasanoff 2006; Levidow and Carr 2009). Innovation in fields of bioethical scrutiny reflects the relationship between democracy and science. It is particularly important, then, to understand how techno-scientific issues (such as biodiversity, GMOs, hygiene, pest control, etc.) are not only configured within technological discourses, but also appropriated in practice, since science and technology *are* political objects that permeate everyday life (Macnaghten 2016).

Grassroots short food supply chains can be understood as efforts to offer alternatives to the current agri-food system – as expressions of responsible innovation. The main idea behind grassroots responsible innovation is that this approach to sustainability goes beyond the idea of mere "product checking" and includes community development, social equity and solidaric transactions. These innovations often focus on food for two reasons: first, because food production is one of the most unsustainable activities on the planet (including issues of energy provision, intellectual property and gross exploitation of humans, animals and natural resources); second, because it is often in the area of food provisioning that communities find the need, the capacity and the margins of action to rethink their participation in the economy.

The following case studies show how alternative provisioning movements in Massachusetts and Lombardy focus differently on the relationship with local communities. Forms of provisioning activism (especially in the area of food, but also in those of commodities, energy and currency) range from the so-called solidarity economy in Italy to CSA schemes in the USA, and include many other actors and organizations. Here I will focus on solidarity purchase groups in Italy and the "local food movement" in Massachusetts as emerging dynamics in which the relationship between producers and consumers is being rethought in multiple, sometimes contradictory, ways. Do short supply chains transform lifestyles? Are these emerging economic circuits promoting environmental, economic and social resilience?

The answer depends on the context, the ideology and the actual practices that follow. On the one hand, the chapters in this book that address the sustainability challenges in food supply chains have an interdisciplinary perspective, and seek to clarify the added value of in-depth qualitative research into short food supply chains. On the other hand, we may encounter different outlooks on responsibility, sustainability and innovation when we discuss these issues, not only in general terms but through the ethnographic observation of relevant local dynamics, as ethnographic research proceeds from the complexity of local contexts rather than from general definitions.

Food communities are particularly suited to the notion of innovation as "the ability to achieve the improbable" (Bassetti 2003). As opposed to an interpretation of *innovation* as a mere technology-fix, Bassetti's intuition is that innovation does not simply mean novelty but rather "something previously improbable, now made real" (Bassetti 2003). He further distinguishes between *capital intensive* innovation and *poiesis intensive* innovation. The latter forges new links between knowledge and agency, which bring into existence previously unknown objects, relations or situations. Innovation thus does not coincide with invention (which is often science and capital intensive), but with the ability to transform lifestyles. In particular, *poiesis* intensive innovation refers to the cultural aspects of work practices, such as beauty and taste, logistics and design. Innovation is thus not seen as an individual creative act, but as a social process that permeates contexts and institutions and, in turn, depends on them.

Poiesis is thus not only the dimension of individual creativity, but the collective and practical capacity of rearranging existing resources in more meaningful ways. It is the dimension of aesthetic and moral appeal instead of mere economic rationality. For example, sustainable tourism, didactic forms of organic farming, culinary typicity, alternative food networks and ecomuseums are all examples of private and public enterprises encouraging *poiesis* intensive innovation in marginal rural areas. Drawing on two personally conducted ethnographic case studies, I will argue that an ethnographic approach to the – often intertwined – ecological, economic, social and political issues implied in innovating the food system favours a focus on creative social experiments and can help us investigate how innovation takes shape in some contexts, but not in others.

Reinventing food chains: experiments in grassroots innovation

My ethnographic observation focuses on Europe (Italy in particular) and the USA (Massachusetts in particular) in comparative perspective. I will argue that grassroots short food chains are an example of *poiesis* intensive innovation, which emerges through a spontaneous aggregation of concerned citizens – both producers and consumers. A rich constellation of grassroots movements are posing the question of food as a political object of risk perception and of collective deliberation, devising and organizing alternative food provisioning networks that value health standards, but also social and environmental sustainability in food production. From Slow Food to CSA, from Solidarity Purchase Groups (*Gruppi di acquisto solidale* or GAS) to Transition Towns, a collective and participatory response to the public perception of techno-scientific innovation and of sustainability is emerging.

In the USA, CSA is now a well-established and commercially viable alternative to the consolidated food industry, especially in urban areas on both coasts. However, working at the high end of a dramatically divided rural–social scenario (dominated by so-called million-dollar farms and very cheap processed food), CSA can more easily speak to affluent sectors of society than can re-engineer social cohesion in local communities, urban health in food deserts and food sustainability in the agri-food system at large. Additionally, CSA is largely producer-driven, a sustainable business model for small or new farmers who wish to target a niche, high-end sector of American urbanites.

The French AMAPs work as pre-paid contracts for the delivery of crops, very much like CSA. Consumers take on the risk of cultivation by paying for crops in advance and thus shield local farmers from risks and bankruptcy; they may also actually participate in crop preparation and harvesting in the field – thus developing a more communitarian feel of attachment to the farm and the farmer, as well as requiring not just money but also time as a (possibly even more precious) form of investment. However, in Europe multiple forms of consumer-driven initiatives have also emerged on a local scale. For example, in Germany the Solidarity Agriculture movement works against turning farms into agribusiness by developing similar partnerships with local farmers.

While solidarity economy networks can be considered as "social networks" involving mainly consumers in networks of self-help, CSA is increasingly seen as a successful business model for neo-rural entrepreneurs (Henderson and Van En 2007). What constructive scenarios do these movements provide for rethinking the role of technology and capital in food production? How do they convey expectations of responsibility in innovation? Is responsible behaviour in food production and provisioning by definition averse to techno-scientific innovation? Comparing two case studies will help to answer this.

1. *The Boston local food movement*

During the period August 2011 to May 2014 I worked in Boston, Massachusetts and had the opportunity to observe and participate in the local food movement that was gaining momentum in the urban area and, indeed, state-wide. This was by way of personal participation in multiple CSA schemes over three years, attending public events and festivals focused on sustainable food in Boston, Cambridge, Somerville, Worcester and Springfield, teaching and supervising research students on Boston University's Master of Liberal Arts in Gastronomy and attending seminars and webinars on the "new economy". The latter included training courses on how to set up a CSA farm, networking events for young entrepreneurs organized by co-working spaces such as the WorkBar in Cambridge, Massachusetts and the FoodLoft in Boston as well as to conferences organized by the Massachusetts state government to encourage "sharing economies" including, but not limited to, food supply. Still recovering from a dramatic economic downturn due to the financial crisis of 2008, governmental, non-governmental and private initiatives were visibly investing – among other "new economy" initiatives – in face-to-face education around food entrepreneurship and technology, such as the Food & Entrepreneurship Meetup in Boston.

As a result, a self-defined "food community" – or a "local food movement" as often referred to at public events – emerged in Boston, encompassing a very wide and diverse range of interests, some of which did not specifically address each other's concerns (if any) with sustainability, some being more focussed on creating new entrepreneurial opportunities, while others addressing equity and social justice through access to healthy *and* affordable food. From encouraging start-up companies selling organic snacks, to CSA schemes offering sliding-scale shares, to increased access to fresh food for less wealthy families and non-governmental organizations (NGOs) openly advocating "food justice", the range of activities and statements was indeed very broad (Loh and Agyeman 2017). Naturally Boston, the cradle of Massachusetts Institute of Technology and Harvard, offers a rich network of mentors, investors, incubators, "accelerators", publishers and working spaces that can tap into the lively political and cultural debates, as well as high incomes in the area, to experiment with short food chains or with sustainable food technology. Examples included Cuppow, a patented drinking lid for preserving jars; Boston Organics, a local food delivery scheme; FoodEx, a sustainable food logistics hub; and Bootstrap Compost, a composting agency offering to pick up food scraps and deliver back compost for gardening. The shared philosophy behind these entrepreneurial efforts is that something creative *and* lucrative can be done to respond to some of the global food challenges such as food waste (about one-third of the food produced across the world is lost or wasted; see FAO 2013). The recipe was intervening in innovative and often tech-driven ways in the production, supply and distribution networks of current food systems, thus impacting urban lifestyles.

The plethora of apps, products, logistics providers and distribution companies that ensue, however, hardly speaks to another pressing problem for contemporary American society: namely, the shocking number of undernourished children and families, the rising epidemic of type-2 diabetes and the large food deserts in which impoverished households without a car and the necessary income cannot find the fresh fruit and vegetables they need to stay healthy (Pringle 2013). Potentially, CSA may be able to respond to this need by ensuring shorter local food chains and putting farmers in touch with the food needs of their surrounding communities. However, this would require that the farmer is a smallholder living in the proximity of these "eaters", producing a diverse range of edible food. Consolidated agriculture in globalized agri-food systems, however, works hectares and hectares of industrial crop-monocultures (typically corn for the food processing industry, soy and alfalfa grass for the production of animal fodder). Additionally, the market logics of demand and supply for fresh, quality food does not necessarily set up a viable conduit between small farmers and malnourished communities.

During the three years I lived in Boston, I became a member of three different CSA schemes, one rather large serving two different pick-up locations in Boston, one much smaller mainly serving central Massachusetts but aiming to expand to the Boston area in partnership with a bike-delivery worker-owned cooperative, and a box-delivery scheme. I also attended farmers' markets both in my immediate neighbourhood and further afield in different locations in Massachusetts. The pressure on farmers was to keep their clients; for example, they introduced loyalty programmes and text message notifications, newsletters, mailing lists, etc. As explained to budding CSA farmers in the webinar organized by Small Farm Central in March 2014,

> there's just more competition for a CSA dollar, and you just have more options than ever to get high quality food. Even our, you know, local … local grocery store … so we're in Pittsburgh, Pennsylvania, I know our … just regular local grocery stores have great produce now. So … there's delivery services, and there's co-ops, there's so many options for people to get really high quality food, so there's more competition for that eating dollar.
>
> (Small Farm Central 2014)

CSA is a marketing device to secure a unique connection between a farmer and a customer *for those who can afford it*. The customer will subscribe to harvest "shares" in advance – usually for a season but possibly for the whole year – so that a higher level of continuity and trust can be established than by occasionally attending farmers' markets. This is the strongest way to support a small farm and it is usually initiated by the farmer through a marketing campaign: distributing brochures, setting up and maintaining a website and email notifications, inviting trusted customers to "potlucks" and events such as apple-picking and hay-making feasts on the farm's premises. As I had occasion to witness, these events are family-oriented, festive and convivial, but strictly based on a commercial relationship between the individual customer-family and the farmer or farm enterprise. Just like visiting a fun fair, kids were entertained by farm workers, video-projections would celebrate the latest projects or farm enlargements, hay-rides would take us into the fields where some explanations would be given on seasonality and types of cultivation; apples would perhaps be picked. However, limited and superficial interaction ensued – as appropriate for a "community" that materializes in the space of a fleeting, however principled and well-meaning, commercial interaction.

Both producers and consumers are aware of the limitations to realizing CSA's potential to reach more disenfranchised sectors of American society. Facilitating schemes, to be fair, were on the rise, including some Massachusetts cooperative credit banks offering loans to students so that they could afford to join a local CSA instead of eating junk foods. Several CSA schemes offered "sliding-scale shares", where one could volunteer to pay a higher price based on self-assessed income so that someone with less income could pay a lower share. However, participating in a CSA scheme remained, on the whole, a costly affair: from the US$2,000 requested for a yearly share in local fruit and vegetables in Boston, to Cape Cod Fish Share schemes to ensure that the New England fishing tradition remains viable and sustainable for both the remaining fishermen and the remaining fish, to the slightly lower prices offered by other farm-share options, there was a clear sense of the privilege enjoyed by those who participated.

2. *Consumer-driven short food chains in Italy*

In Italy, solidarity purchase groups (*gruppi di acquisto solidale* or GAS) have emerged slowly since the 1990s as networks of consumers who work in close collaboration with individual producers (or sometimes cooperatives) to procure food and other items as far as possible directly, on a local basis and in a sustainable manner. By seeking a direct relationship with producers and cooperatives, these groups move beyond ethical consumption towards engaged and collective forms of procurement. The sociologist Michele Micheletti (2003) has defined political consumerism as "individualised collective action", namely, an individual's personal involvement in a wider movement of critical consumption by way of boycotting or buycotting. Micheletti argues that critical consumption networks provide ways of assuming individual responsibility by moving beyond the role of mere consumer, triggering new lifestyles and thus turning food into a political object. Furthermore, with GAS food becomes political because consumers organise themselves as *collectives*, in order to exercise a form of political agency in the economic system.

My case study is based on two years when I participated in a GAS collective in Bergamo in northern Italy, and on my work as a delegate of my group in a network of 32 GAS in the province of Bergamo (an area of about one million inhabitants comparable to that of Greater Boston) in the period spanning 2009 to 2011. GAS adhere to a philosophy of direct relationship with their providers, who they call "producers" to indicate that they wish to procure their food directly from growers. Their National Charter promotes "respect for the environment and solidarity between the members of the group, the traders and the producers" (Gruppi di Acquisto Solidale 1999). They are groups of, on average, 20 to 40 families who directly manage the logistics of collective purchase: this work is largely based on voluntary and informal self-organization.

In practice, GAS members convene periodically (usually once a month, though this varies greatly according to local preferences and to group size) to organize and share the tasks of bulk-buying from trusted producers. These may be local organic farmers or breeders (for vegetables, fruit and meat), dairies for milk and cheese, or preferably cooperative or farmers' consortia as processers of oil and basic foodstuffs such as pasta, sauces and fruit juices. The national network of GAS, *Retegas,* estimates that about 100,000 people practice this method of provisioning. Their collective practices include assessing the producers, organizing on-site visits and holding social events. In many cases, participating in a group means being socialized into home bread making, cherry picking, grape harvesting, chestnut picking, etc. I propose to look at all these practices as "cross-cutting counter-epistemologies" (Grasseni 2013), namely, as overlapping practices and routines that develop new competences. These are notably not necessarily farming skills, but rather logistics and accounting; reciprocity and leadership within the group; information processing and communication about organic certification and food labelling.

Among the examples of responsible innovation adopted by GAS are the capacity to "network up" among several groups. These larger networks may contribute to very local and specific forms of responsible innovation – for example, engaging local administrations to provide them with a public space for managing orders or lobbying for the provisioning of organic food to school canteens. "Districts and networks of solidarity economies" – DES (*distrettidi economia solidale*) and RES (*reti di economia solidale*) – are examples of responsible innovation through which networks of GAS establish alliances with other "alternative" socio-economic actors such as Fair Trade shops, ethical banks, responsible tourism agencies or environmental associations to develop specific projects. For example one such alliance promoted the use of local wheat *cultivars* to make bread. GAS are a pivotal point in these short chains as they provide the necessary economic guarantee that the product will have buyers.

Across these networks, food works as a social aggregator through practice and knowledge. These very localized and concrete, day-to-day practices have the ambition of trying to respond to global issues such as fair price and food sovereignty, or that of moral solidarity between producers and consumers. GAS implement a sort of practical moral philosophy: responsible behaviour in food provisioning is not about making *big* normative choices, but rather about choosing an innovative course of action with soft technologies and a laborious relational framework which may change one's lifestyle – a *poiesis*-intensive form of innovation.

The range of examples of experiments with food as a form of responsible innovation is in fact widespread and diverse. Sarah Pink describes the similarities between the philosophy of the Slow Food movement and that of the *Cittaslow* (slow city) network. Both wish to engage their members in "educational" projects in a broad sense: Slow Food is interested in taste education, *Cittaslow* in quality living. The *Cittaslow* charter includes measures such as places for people to "sit down and rest, not just in town centres" or plans to "plant sweet smelling or environment-enhancing plants in public and private gardens" (Pink 2009). This example works at a different scale than the entirely spontaneous networks of consumer-citizens in GAS, as "Cittàslow inevitably begins life as a top-down process, in that it is necessarily led by the town council" (Pink 2009). *Cittaslow* activists are not grassroots militants but rather exercise their influence through participating in governance: Pink sees *Cittaslow* as "working as a form of persuasion by example" (Pink 2009). Sustainable local economy is, for instance, one target of *Cittaslow's* policies, and Pink mentions the rationale of *Cittaslow* administrators in acting against encroaching global supermarket chains. *Cittaslow* does not campaign directly against Tesco or other chains, but rather invests in creating closer links between producers and consumers, such as through farmers' markets. In contrast, GAS get together to buy directly from producers, thus promoting knowledge of their territory and of its seasonality.

For both, the aversion to supermarkets is motivated by the correlated disappearance of small shops for local customers (especially elderly people with no car), with the effect of driving most economic transactions out of historical town centres to purpose-built shopping malls. Additionally, small producers are often bound to distribution chains by contracts that set very low prices while binding them to ensuring fixed amounts of quality products by a deadline. This guarantees producers that their produce is going to be bought, but it also drains it, at cheap prices, from local markets where it could be sold for more during the high season. This kind of double-bind with chain distributors has to be taken into consideration when proposing alternative solutions such as farmers' markets. If new social partnerships are to support regional economies, the governance of these processes must be coupled with grassroots forms of responsible innovation. The contribution of the GAS movement in Italy seems precisely to be that it has introduced new forms of engagement (Forno and Graziano 2014), including practices that can "trickle upwards" into the institutional and administrative fabric.

Conclusion

A wide range of models of sustainable food production and consumption are being experimented with, from high-tech, capital intensive entrepreneurship to *poiesis*-intensive forms of consumer self-organization in collective food procurement networks. I considered grassroots short food supply chains as possible examples of *responsible innovation* as they re-invent basic, everyday, universal practices such as that of food procurement in innovative ways. These strategies, however novel they may look, should be considered against their diverse socio-economic backgrounds and the political experiences that have preceded them. CSA schemes are, in a sense, contemporary re editions of initiatives that started in the 1960s and 1970s with the first American consumer cooperatives. However, scholarship on alternative agri-food networks has highlighted that these initiatives often focus on quality and sustainability, and less on social inclusion and food justice – namely, access to resources for underprivileged groups (Mares 2014; Luetchford 2016; Alkon and Guthman 2017).

In Italy, with active political support, especially from left-wing parties, some of these initiatives have developed into nationwide supermarket chains such as Coop, a large distribution "cooperative". Slow Food – often critiqued for being elitist in its price-blind pursuit of quality – stems from Arcigola, the ex-Communist after-work association that offered popular cultural activities in the spheres of music, gastronomy, excursions, etc. GAS emerged as a form of consumer self-organization that preceded the widespread crisis in household incomes and food consumption in Italy. There is no obvious connection between the development of the GAS networks in Italy and the economic crisis that began in 2008, became particularly intense in 2011, and peaked during 2012–2013. However, during the crisis years GAS began to attract consistent attention from the media and even political parties, as they seemed to represent a bottom-up form of grassroots innovation in the name of solidarity and resilience. It is, in any case, mostly white, middle-aged, middle-class people who can afford the money, and especially the time, to make self-conscious choices about their diet and food procurement practices.

Community participation takes varying forms according to the dominant discourse in the respective contexts. While innovative entrepreneurship is invoked in the Boston area as part of a local thriving knowledge economy, Italian consumers mobilize strategies of self-rediscovery as communities of heritage, of conviviality and of neighbourhood, through novel readings of local landscapes and the re-evaluation of their agricultural vocation vis-à-vis expanding supermarket chains. This is particularly apparent in the small and medium towns adhering to *Cittaslow*, where the administrative and economic fabric of society experiments with food reinvention through a heritage framework (Grasseni 2017). These processes include the rise of new types of partnerships involving producers and consumers, administrators and citizens in local practices of economic and institutional co-production (such as *Cittaslow* or so-called districts of solidarity economy). Their analysis introduces an context-oriented variety to the debate about responsible innovation, which requires ethnographic observation and the cultural study of contexts to appreciate how differently ideas of responsibility, sustainability and innovation actually synergize within localized discourse and the practices of citizenship, trust and knowledge as well as resources.

In a comparative perspective, the limits and potential of each of these phenomena become evident. Alternative food networks still have to demonstrate a tangible impact upon local economies and improvement in sustainable lifestyles, while they are usually voluntary-based, time-consuming modes of engagement. They have so far instead showed a great capacity for self-organization and for implementing self-organized forms of lifelong education in food production and network management. Precisely because of their successful but socially particular proliferation, it is worth teasing out the strategies, motivations and critical points they share with other movements and initiatives. In this chapter, I proposed that an ethnographic insight into responsible innovation can assist an understanding of the diverse modalities and added value of localizing agri-food systems. Short food supply chains disseminate responsible innovation in the form of new partnerships, practices and networks –for example, circuits of peer-to-peer help or co-produced solutions to logistics, management and organization. Though not originating from the economic crisis, theses type of social networks facilitate the application of innovative alternatives in new, challenging contexts, *also* using the economic crisis to propose pathways and lifestyles that focus on solidarity and the environment as a common good.

Acknowledgement

Fieldwork was partly funded by the Wenner-Gren Foundation for Anthropological Research (Seeds of Trust: A Comparative Analysis of Solidarity Economy Networks in Lombardy (Italy) and Massachusetts (USA), 2013/14). Research continues in the framework of the project *Food citizens? Collective food procurement in European cities: solidarity and diversity, skills and scale*. This project receives funding from the European Research Council under the European Union's Horizon 2020 research and innovation programme (grant agreement No. 724151).

References

Abrahamson, S. and Bertoni, F. (2014). Compost Politics: Experimenting with Togetherness in Vermicomposting. *Environmental Humanities* 4(1): 125–148.

Alkon, A. and Guthman, J. (2017). *The New Food Activism: Opposition, Cooperation, and Collective Action*. Berkeley, CA: University of California Press.

Bassetti, P. 2003. *Innovation, Social Risk and Political Responsibility*. Lecture presented at the London School of Economics, 14 May 2003. Available at: www.fondazione-bassetti.org/en/pages/2009/04/innovation_social_risk_and_pol.html (accessed 4 March 2017).

Bellarby, J., Foereid, B., Hastings, A. and Smith, P. (2008). Cool farming: Climate impacts of agriculture and mitigation potential. Greenpeace International.

Faigel, J. and Sinfield. J.V. (2017). Scaling Impact: Equity, Access, and Opportunity in Boston's Regional Food Economy. *Social Innovations Journal* 41, 4 December 2017. Available at: www.socialinnovationsjournal.org/editions/issue-41/75-disruptive innovations/2680-scaling-impact-equity-access-and-opportunity-in-boston-s-regional food-economy (accessed 8 May 2018).

FAO (2013). Food wastage footprint: Impacts on natural resources. Summary report. Available at: www.fao.org/docrep/018/i3347e/i3347e.pdf (accessed 21 August 2018).

Forno, F. and Graziano, P. (2014). Sustainable Community Movement Organizations. *Journal of Consumer Culture* 14(2): 139–157

Grasseni, C. (2013). *Beyond Alternative Food Networks. Italy's Solidarity Purchase Groups*. London: Bloomsbury.

Grasseni, C. (2017). *The Heritage Arena: Reinventing Cheese in the Italian Alps*. New York: Berghahn Books.

Gruppi di Acquisto Solidale (1999). Documento base dei GAS. Un modo diverso di fare la spesa. Available at: www.retegas.org/upload/dl/doc/GASDocumentoBase.PDF (accessed 4 March 2017).

Hankins, J. (2019). Grassroots Case Studies in Poiesis Intensive Responsible Innovation. In: R. von Schomberg and J. Hankins (eds.), *The Handbook of Responsible Innovation: A Global Resource*. London: Edward Elgar Publishing.

Henderson, E. and Van En, R. (2007). *Sharing the Harvest: A Citizen's Guide to Community Supported Agriculture*. Burlington, VT: Chelsea Green Publishing.

Henriques, A. and Richardson, J. (eds.) (2004). *The Triple Bottom Line: Does It All Add Up?* Abingdon, UK: Earthscan.

Jasanoff, S. (2006). *States of Knowledge: The Co-Production of Science and the Social Order*. London: Routledge.

Krause, E. and Sharma, A. (2012). Sustainability "Wars" in a New England Town. *Futures* 44: 631–641.

Levidow, L. and Carr, S. (2009). *GM Food on Trial: Testing European Democracy*. London: Routledge.

Loh, P. and Agyeman, J. (2017). Boston's Emerging Food Solidarity Economy. In: A. Alkon and J. Guthman (eds.), *The New Food Activism*. Berkeley, CA: University of California Press, pp. 257–283.

Luetchford, P. (2016). Ethical Consumption: The Moralities and Politics of Food. In: J. Klein and J. Watson (eds.), *Anthropology and Food. A Handbook*. London: Bloomsbury, pp. 387–405.

Macnaghten, P. (2016). Responsible Innovation and the Reshaping of Existing Technological Trajectories: The Hard Case of Genetically Modified Crops. *Journal of Responsible Innovation* 3(3): 282–289.

Mares, T. (2014). Engaging Latino Immigrants in Seattle Food Activism. In: V. Siniscalchi and C. Counihan (eds.), *Food Activism: Agency, Democracy, and Economy*. London: Bloomsbury, pp. 31–46.

Micheletti, M. (2003). *Political Virtue and Shopping. Individuals, Consumerism and Collective Action*. New York: Palgrave Macmillan.

Mol, A. (2009). Good Taste. The Embodied Normativity of the Consumer-Citizen. *Journal of Cultural Economy* 2(3): 269–283.

Morgan, K., Marsden, T. and Murdoch, J. (2006). *Worlds of Food. Place, Power and Provenance in the Food Chain.* Oxford: Oxford University Press.

Pink, S. (2009). Urban Social Movements and Small Places. Slow Cities as Sites of Activism. *City* 13(4): 451–465.

Pringle, P. (ed.) (2013). *A Place at the Table: The Crisis of 49 Million Hungry Americans and How to Solve It.* New York: Participant Media and Public Affairs.

Small Farm Central (2014). Webinar organized by Small Farm Central on community supported agriculture, 6 March 2014.

Ubois, J. (2011). Responsible innovation/sustainable innovation. Fondazione Giannino Bassetti. Available at: http://issuu.com/fondazionebassetti/docs/jeff-innovation-aaa-2010-2 (accessed 21 August 2018).

4 Food localization and agency

The cases of Regionalwert AG and Luzernenhof in Freiburg, Germany

Lena Partzsch

Introduction

Food localization is a challenge that is being tackled by many local initiatives around the world. Their objective is to grow (or raise) and harvest food close to consumers' homes, and then to distribute it over much shorter distances than is common in the global industrial agri-food system. This chapter reflects upon agency and accountability in such initiatives. Central models of change attach great significance to individual agents for pushing disruptive change (Moore *et al.*, 2012; Westley *et al.*, 2007). So-called "social entrepreneurs" are the inventors of a novel norm, idea or product (Moore and Westley, 2011, p. 5). In the area of Freiburg in Breisgau, Germany, Christian Hiss serves as an illustrative example of such an agent. He has received several international awards, including Ashoka and Schwab Foundation fellowships, for his innovative Regionalwert AG enterprise. The name literally means "regional value plc" and is a "citizen" stock company that invests in local agri-food businesses (Regionalwert AG, 2016). Agency is enabled here through the market and higher prices paid for local and organic food products. In contrast is the competing alternative project Luzernenhof (*Hof* means farm; literally "Lucerne's farm") in the area of Freiburg. Luzernenhof is organized as a community-supported agriculture (CSA) project that experiments with practices beyond the market. The aim of my chapter is to show that, while food localization is often portrayed as "best practice", it also implies decisions about alternatives and this raises questions of agency and accountability.

A major challenge for food localization around Freiburg is farm dieback. Only 10 per cent of heirs continue family farming in the region (Blum, 2010). Moreover, due to high land prices, it is difficult for young farmers to start their own businesses. Regionalwert AG provides capital for start-ups and successions dedicated to socially and ecologically sustainable farming (according to the EU- Eco-Regulation). Shareholders accept a marginal or zero dividend; borrowers repay loans with below average interest rates over long periods of time (Bognanni, 2010). By 2015, Regionalwert AG had invested 2.2 million euro in 18 organic businesses (Hodeige, 2015). In the case of Luzernenhof, when the original owner, Marga Widmaier, wanted to retire, she turned the farm into a CSA project (Luzernenhof, 2016). In CSA projects, consumers cover the costs of farming rather than paying a particular price for a specific food product (see Chapter 11). So there is no need for a farm to be profitable and to compete on the conventional (organic) market.

In the second section of this chapter, I outline Freiburg' agri-food background. Conservative and green values have been crucial in preserving regional farming in the area. In section 3, I review the relevance of social entrepreneurs as key agents of change in transition processes, based on the *social innovation cycle* (Moore and Westley, 2011; Westley *et al.*, 2011). I reflect upon the assumptions made in this model with reference to the structure–agency debate (Hayward and Lukes, 2008). In particular, I outline links between agency and accountability. In the third and fourth sections, I analyse how the two cases of Regionalwert AG and Luzernenhof are enabling and/or constraining agency. Methodologically, my analysis is based on the two projects' self-portrayals on the home pages of their websites, their publications (newsletters, meeting protocols, books, etc.), media interviews they have given and newspaper articles about them, participatory observations at promotional events, several field visits and ten semi-structured interviews with participants (section 4).

The final section compares and discusses agency and accountability in the two projects. Bringing together these two alternative cases, I am able to contribute to a more comprehensive understanding and critical reading of agency and accountability in processes of food localization. Both initiatives allow for new agency beyond the global industrial agri-food system. At present, Luzernenhof is enabling farmers' agency much more than Regionalwert AG. It permits young people to become farmers without getting into debt and without paying high rents for agricultural land. However, the farm completely relies on voluntary contributions, donations and unconditional loans. By contrast, in Regionalwert AG, farmers have to compete in and are constrained by the conventional (organic) market. In contrast to Luzernenhof, it is not at the mercy of charity but demands "fair prices" based on a transparent accounting system.

Freiburg: organic pioneer city

Freiburg is a city in the prosperous Black Forest area in the south-west of Germany with a population of around 220,000 people. Traditional food and beverages include spätzle, a kind of soft egg noodle, Black Forest cake and Riesling wines. The city of Freiburg is particularly known as a "green city" in Germany. It has received many environmental city awards, ranging from Environmental Capital of Germany (1992) to a Freiburg presentation on City Leadership Day at the UN Rio Summit in 2012 (Frey, 2011; Freytag *et al.*, 2014). The first Green Party mayor in Germany was elected for the city of Freiburg in 2002. Moreover, the city is home to a large Christian community, founded by Rudolf Steiner who defined the rules for "biodynamic" agriculture. Demeter International, which controls these rules, is a leading organic certifier worldwide. Both case studies, Hiss's farm in Regionalwert AG and Luzernenhof, follow the Demeter standard.

While studies have highlighted the pioneering role of Freiburg in green alternative developments, they have generally been inattentive to conflicts over and among specific "green" alternatives. In particular, when emphasizing the importance of highly respected civic leaders among Freiburg's alternative movement (e.g. Buehler and Pucher, 2011; Frey, 2011), scholars often lose sight of people with more progressive ideas who could not prevail against the mainstream. The two case studies analysed below represent two possible alternatives in food localization in and around Freiburg.

Regionalwert AG has existed since 2006 with the administrative office in the small town of Eichstetten, 18 km (11 miles) outside of Freiburg. This is where the first partner company, Hiss's own farm, the Demeter market garden Querbeet (literally "across the bed") is based (Siebold, 2011). As all 18 partner companies currently come from the region of Freiburg, a regional network has grown over the last ten years that allows companies to intensively trade with each other and to successfully market their products together (Regionalwert AG, 2016).

The second case of Luzernenhof is a CSA project located in the village centre of Seefelden, 26 km (16 miles) outside of Freiburg (Luzernenhof, 2016). A team around the farmer couple, Magdalena and Johannes Supenkämper, has been running the once family-owned farm of Luzernenhof since 2012; it includes cattle breeding, a cheese dairy, vegetable gardening and beekeeping. In terms of food localization and disentanglement from the global market, Luzernenhof has a more ambitious agenda compared to Regionalwert AG (Luzernenhof, 2016). As my analysis below will show, this implies different kinds of agency and of accountability, respectively. In the following section, I will first discuss the concepts of agency and accountability in theoretical terms and develop an analytical framework, before I analyse the two empirical cases in sections four and five.

Agency in the social innovation cycle

Central models of change attach great significance to individual agents in initiating disruptive change (Finnemore and Sikkink, 1998; Westley *et al.*, 2011), and a growing body of literature suggests that individual entrepreneurs are also crucial in environmental governance (Boasson and Wettestad, 2014; Jordan and Huitema, 2014) and establishing decentralized agri-food systems (Shey and Belis, 2013). Social innovations need to fit into the context of established systems, and it takes entrepreneurs who strategically frame innovations, such as food localization, to give them sense in the surrounding normative environment (Finnemore and Sikkink, 1998; Mintrom and Norman, 2009).

This chapter aims to explore innovations regarding food localization, i.e. the decentralization of agri-food systems, through two initiatives on the ground. Westley *et al.* (2007, pp. 65–68) suggest a model that can help us to gain an understanding of such transitions (see also Moore *et al.*, 2012; Westley *et al.*, 2011). Their *social innovation cycle* is based on Holling's (1978) adaptive cycle of resilience research and distinguishes four phases organized as a back and front loop process: (1) release, (2) reorganization, (3) exploitation and (4) conservation (see Figure 4.1).

In the first phase of the change process, *release*, problems need to be identified and set on the agenda before powerful rules and institutions collapse (Moore and Westley, 2011; Westley *et al.*, 2007). For instance, carbon emissions caused by the established agri-food system: someone has to address this problem, before the decentralized supply system can replace the conventional agri-food system of long distances between food producers and consumers.

The second phase of the change process, *reorganization*, consists of selecting the (supposedly) best option to address the problem; for instance, support for regional agri-food businesses which make long-distance transport obsolete and hence cause fewer carbon emissions. This phase signifies a precarious moment from the point of view of whether the affected system remains in its current basis of stability, adapting and learning but not transforming, or whether it is pushed close to a threshold that tips the system into a new stable domain. It is only in this second phase of the innovation cycle that the diffusion of an invention is possible, either through transformative or incremental change (Westley *et al.*, 2011).

Westley *et al.* (2011, p. 771) follow an actor-centred approach and argue that "[s]ystematic innovation strategies [...] cannot depend purely on market forces, nor can they deliberately be planned. [A]gency clearly plays a role at each stage of the process". In particular, in the first phase of release, so called "social entrepreneurs" who are "the inventor[s] of a novel norm, idea, or product" (Moore and Westley, 2011, p. 5) are crucial for the *social innovation*

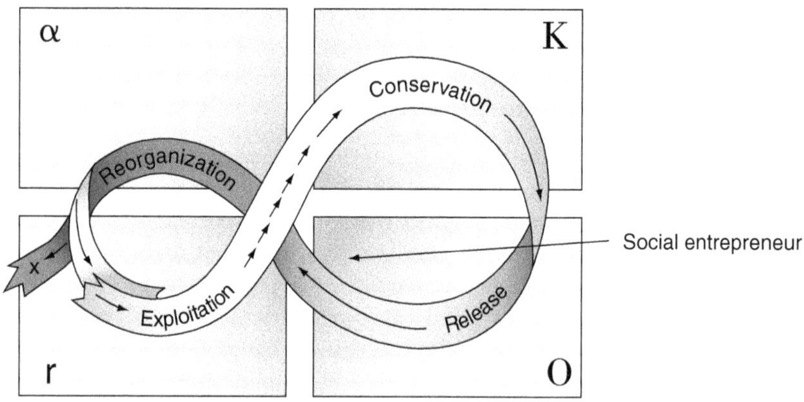

Figure 4.1 Holling's 4-Phase Adaptive Renewal Cycle (1986).
Source: author's illustration based on Resilience Alliance (2016).

cycle. Grenier (2008) outlines that social entrepreneurship provides agency in a globalized world. Agency in this context is understood as *self*-determined and independent from social constraints.

In this vein, Pitkin (1985, pp. 276–279) emphasizes how an actor may have the *power to* accomplish something all by him or herself, assuming that self-determined agency beyond structural constraints is possible. By contrast, constructivist and post-structuralist research has demonstrated, and continues to demonstrate, how every actor or group is defined through socially constituted relations that at least indirectly shape the actions of individuals (Barnett and Duvall, 2005; Wendt, 1992). This means that structures define who the actors are and what they are able to do.

The debate on *subjectivity* is concerned with the externally determined constructions of identities ("me" and "us" versus "the other") and how agents do (not) reproduce systems and positions. Butler (2004) emphasizes how the subject is only produced by the very act of speaking and acting in its name. While exploring the problematic of subjectivity and vulnerability, post-structuralist scholars disclose individual hero narratives (Zehfuss, 2007). From this perspective, the social entrepreneurship discourse itself has merely produced individual "change-makers". In other words, Christian Hiss only became a social entrepreneur through respective awards and discourses. He did what he did to perform his social role. This difference – whether we assume Hiss to be a self-determined agent or subject to structural dynamics – has far reaching consequences with regard to his accountability (Hayward and Lukes, 2008). If we assume the ability of individuals and groups to get things done on their own, there also is an individual accountability for non-action. Gridlock can be explained by structural constraints *and* a lack of "new thinking" (Partzsch, 2017).

At the nexus of philanthropy and its growing popularity through awards and prizes, in particular, as well as with secondary research on them, social entrepreneurs have become increasingly present in social transition theory and practice (Mair *et al.*, 2006). In expert commissions, they often represent the local context (Grenier, 2008). If we admit that structure and agency are not mutually exclusive (Hayward and Lukes, 2008), we should re-examine the significance of individuals in food localization against the backdrop of the structure–agency debate. Assuming that "new thinking" is possible and that it takes self-determined agency for transformational change, we should first ask "if" and, if yes, "when and how" (individual and collective) agents matter. Second, we should analyse how transition processes such as food localization either create new additional and/or prevent prospects of self-determined agency. In this chapter, I would like to link these two questions to questions of accountability. If we know who is able to exercise self-determined agency, we are also able to hold these people accountable for their (non-)action. In the following case studies, I analyse and compare these aspects for the two cases of Regionalwert AG and Luzernenhof.

Case-studies

Regionalwert AG: a new accounting system to promote food localization

> I am not concerned with the public interest. Regionalwert AG is no charity but a business company. Money is crucial.
>
> (Christian Hiss, cited in Bognanni, 2010)[1]

The case of Regionalwert AG very well illustrates the *social innovation cycle*. In the first phase of release, Christian Hiss identified problems in the conventional food system as well as in organic agriculture. Hiss's father founded one of the first German organic farms in Eichstetten in 1952 (Prummer, 2011; Steckmeister, 2012). After completing his vocational training as a gardener, Hiss established his own successful organic nursery, Querbeet, nearby. His Demeter products have met the high demands of ecologically and socially conscious communities in and around Freiburg. For example, Querbeet has established weekly vegetable markets at Freiburg's Rudolf Steiner (Waldorf) schools. However, when Hiss wanted to expand the Querbeet farm in 2000, he could not obtain a bank loan (Prummer, 2011).

Hiss came to the realization that his profit margins were too small, especially compared to conventional farms which externalize social and environmental costs, such as nitrate pollution of surface and ground water, through chemical fertilizers (Prummer, 2011). Farmers who want to expand or start new organic businesses do not get bank loans due to their profit margins being perceived as inadequate (Blum, 2010). Besides farming organically and conserving farms in the region, Hiss wanted to farm profitably: "I did not say, f*** capitalism, and give up; instead, I learnt how to work with its instruments"[2] (cited in Prummer, 2011). In 2006, Hiss transformed his farm into a stock company, Regionalwert AG, and selling shares allowed him to finally expand his business.

Hiss's selling point vis-à-vis potential investors is that they also benefit in ideational terms from preventing further farm dieback and from a local, fresh and healthy food supply (Prummer, 2011). Indeed, 75 per cent of his investors come from the region of Freiburg (Steckmeister, 2012). They bear a high financial risk and expect little (or no) return. Furthermore, the shares are not tradable, and a single shareholder can possess a maximum of 20 per cent of the shares (Blum, 2010). Hiss himself is chairperson of the Regionalwert AG board (Regionalwert AG, 2016). He convinced 500 shareholders – "citizens" – to invest 2.24 million euro into his regional "citizen" stock company (Schindler, 2014).

Following investments in Hiss's own farm, Regionalwert AG has continued to invest in organic farms and other companies further down the food supply chain in the region, such as caterers and organic food stores – either as silent, minority or majority shareholders. Regionalwert AG's definition of "organic" is based on the EU-Eco-Regulation (which is the minimum standard of organic farming). Hiss started to rent out the nursery, Querbeet, to a younger farmer, who told me that he would otherwise not have been able to cover the initial costs of his own farm business (interview on 28 June 2016). Another example of a partner company is Zwergenkueche (literally "dwarfs' kitchen") which provides freshly prepared, organic meals to kindergartens in the village of Endingen (30 km, or 19 miles, from Freiburg). Maren Huth and Sabine Joseph founded this company in 2013. As a mother and a kindergarten teacher, the two women had recognized the poor quality of food provided to children in kindergartens, and Regionalwert AG allowed them to offer a sustainable alternative (field visit on 28 June 2016). At the same time, it is possible that, within the Regionalwert AG network, Zwergenkueche processes food that farmers are not able to sell. So the inclusion of food processing businesses, such as Zwergenkueche, allows Regionalwert AG to avoid food waste and reduce the risks of overproduction. The partner companies are and need to be profitable, otherwise the investors lose money.

In order to understand the full value of Regionalwert AG, Hiss started a double accounting system. In addition to conventional profit and loss accounting, Hiss has recorded transport distances of the products, the kind of salaries paid, whether employees have permanent contracts or are seasonal workers, the origin of manure nitrogen and how soil fertility and cultural landscape develop (Blum, 2010). These indicators pave the way for an internalization of the environmental and social costs of farming and the valorization of nature. Hiss envisions a market model similar to carbon emission markets, in which farmers trade pollution rights and in which organic farming companies like Regionalwert AG will be in a good position to financially benefit from their efforts to decentralize agriculture (Blum, 2010).

In sum, after having identified a particular problem (lack of investment for local organic agriculture due to inadequate margins) in the phase of release, Hiss created a viable alternative to solve the identified problem in the phase of reorganization (the "citizen" stock company which develops a new accounting system). People in Hiss's personal environment agreed that Regionalwert AG is his "baby" and that he as an individual rightfully received awards, such as Ashoka and Schwab Foundation fellowships (field visit on 28 June 2016; interview on 2 September 2016).

Luzernenhof: food localization beyond the market

> This is not merely a matter of paying for milk, yogurt and cheese; this is about supporting an idea, and this is sometimes difficult to convey.
>
> (Luzernenhof farmer, cited in Hardt, 2012)[3]

Agency and accountability are less clear in the case of Luzernenhof. In 1986, Marga and Joachim Widmaier converted the family farm of Luzernenhof, which they ran in its fourth generation, to biodynamic agriculture (Stuttgarter

Zeitung, 2013). In the phase of release, similar to Hiss, the Widmaiers realized that biodynamic agriculture is very expensive, and selling their farm products on the market did not allow them to run the farm in an economically sustainable manner. (Hardt, 2012; Saurer, 2012). When the Widmaiers wanted to retire, it was impossible to find successors (interview 6 September 2016).

In particular, Luzernenhof is based in a village, as many cattle farms traditionally used to be in the region. This means that the Widmaiers had to laboriously transport hay from the fields to the cowshed in the village (notes from field visit 30 July 2016). Although they produced milk following the Demeter rules of biodynamic agriculture, the Widmaiers were unable to introduce their milk into the organic market. The organic milk float of Breisgau Milch, the only regional collector, simply did not pass by their farm (interview 6 September 2016). Therefore, the Widmaiers started to produce cheese and distribute it through organic food shops; however, for the bulk of the dairy, they received only the low price of conventional milk (interview 6 September 2016).

When they wanted to retire, the Widmaiers first turned to Regionalwert AG and decided, in 2010, to become part of the network. Marga Widmaier sold land to Hiss, and the "fresh" capital allowed her to hire a farmer couple to run the farm (while she had not yet retired). However, for various reasons, the couple did not even stay for a full season (interview 6 September 2016). The Widmaiers lost a lot of money, and they realized that the value of preserving an organic cowshed farm in the village of Seefelden was not possible within the conventional (organic) market (notes from field visit 30 July 2016).

In 2011, the Widmaiers transformed Luzernenhof into a CSA project. In practical terms, this move allowed the Widmaiers to attract people to work on their farm in exchange for free board and lodging. In legal terms, they created a private corporation, in which Marga Widmaier was an equal shareholder among most other farm team members (interview 6 September 2016). As Luzernenhof was only a dairy farm at the time, the team mainly approached members of another CSA project close to Freiburg, which was focused on vegetables, to add dairy to their members' nutrition. When this plan did not work out, Luzernenhof started to grow vegetables in addition to dairy production (interview 6 September 2016). Providing all basic food supplies from only one farm, "like in the good old times", has become an essential feature of Luzernenhof's communication strategy (field visit 30 July 2016). In 2016, the farm was producing food for 150–200 people (Luzernenhof, 2016).

Luzernenhof's annual budget is around 240,000 euro, and each CSA member would need to pay an average of 130 euro per month to cover this budget (Hardt, 2012). However, there is no relation between an individual contribution and the amount of food received, and members also bear the common risk of poor harvests and mismanagement. The amount of the membership fee is not fixed. CSA members always make pledges at the end of each year for the upcoming season. Some pay considerably less than 130 euro per month; children are nourished without making individual pledges (interview 5 September 2016).

While, in theory, CSA members are supposed to be able to live completely on the farm's products, the Luzernenhof team has never assessed how much food members buy in addition to what they receive from the farm. Members told me that this varies a lot (interview 6 September 2016). Sometimes members (or their children) simply do not like the taste of the cheese and therefore buy different cheese in the supermarket (notes from field visit on 30 July 2016). As members have to make their food requests in advance at the beginning of each week, they also face situations of higher demand that they did not anticipate. In such cases, they even buy milk in the supermarket. At the same time, Luzernenhof frequently overproduces milk (the cows continue producing milk independently from the members' demands) (interview 6 September 2016).

As the farm team was not able to buy out the Widmaiers, who wanted to retire, based on membership contributions in 2012, a private individual gave an interest-free loan to the farm team to buy the land, and two people provided another interest-free loan to purchase the farm buildings. Who these lenders are is a secret, and only a few people working on the farm know their names. The lenders inherited a fortune and invested in the farm, although they are not CSA members themselves. Land registry entries serve as collateral for loans (interview 5 September 2016).

As of September 2016, the farm team had 10 members of which only the three male professional farmers were shareholders of the Luzernenhof corporation. They were also the only workers on the farm who the corporation employs on the condition that they are subject to German social security deductions (including retirement payments, health insurance, unemployment insurance, etc.). The other seven members of the farm team, including Magdalena Supenkämper who is a professional farmer, worked on a voluntary or marginal employment basis (interview 5 September 2016). There is a high turnover among people working on the farm (interview 5 September 2016). For this reason, it is virtually impossible for CSA members to hold members of the farm team accountable for mismanagement (often non-decisions). Moreover, they lack expertise and knowledge about the farm's practices. "In theory, you can ask questions at the annual meetings, of course. However, you just sit there. You trust Johannes [one of the three male farmers]. And you feel bad, as you can't make higher pledges"[4] (CSA participant, interview 31 August 2016).

The farm team's living standards completely depend on CSA members' voluntary pledges, on additional donations and on the two secret people who provided the unconditional loans to pay out the Widmaiers. While the Widmaiers took agency when transforming the farm into a CSA project, they have now completely left the farm to the CSA team. There is a high turnover in the farm team, and members often lack information. Agents' interactions are hence trust-based, and the survival of Luzernenhof depends on charity.

Discussion

The lack of a market or sustainable business model defines the central difference between Luzernenhof and Regionalwert AG. In both cases, farmers had to deal with the issue of farm dieback that they agreed to be the result of margins being too small in decentralized and organic agriculture. Both initiatives faced structural constraints in the conventional (organic) food market. However, while Hiss was relatively well embedded in the alternative "green city" structures of Freiburg (vegetable markets at Steiner schools, etc.), Luzernenhof was not even on the route of the organic milk collector. So, while Hiss had only to be innovative in terms of finding investors, Luzernenhof had to transform more fundamentally in response to market constraints.

Hiss became a "change agent" by creating the "citizen" stock company. "Citizen" shareholders can hold him accountable for his actions within Regionalwert AG. Moreover, partner companies and consumers have relations clearly defined through the market. Partner companies are accountable to the shareholders of Regionalwert AG. Consumers can hold producers accountable by either buying or not buying their local and organic products at above average prices. While Hiss's successor at Querbeet told me that Regionalwert AG provides him with the option to leave the profession of farmer at any time (in contrast to the conventional model of a family farm in which heirs are forced to stay in the farm business in order to preserve local agriculture), another partner farmer said that the requirement to repay his debt does not leave him much choice and forces him to constantly work overtime (field visit on 28 June 2016).

All the Regionalwert AG farmers I spoke to, moreover, emphasized the significance of land ownership (field visit on 28 June). A substantial share of their production costs includes the rent for agricultural land. So while the "citizen" shareholders renounce dividends or interest on their capital investments, land owners insist on rent for their land. They financially benefit from the farmers' work and the consumers' willingness to pay higher prices for food. As there is a very high demand for land in the region of Freiburg, organic farmers in the "green city" region are not usually worried about finding consumers willing to pay higher prices for organic products, but about land owners willing to rent their land, especially with regard to the biofuel (maize hybrid) boom and respective producers able to pay higher rents. In this situation, the Regionalwert AG provides agency to these organic farmers though additional capital.

Compared to Regionalwert AG, agency and accountability are less definite in the case of Luzernenhof. The Widmaiers, who transformed the farm into a CSA project, did so to be able to retire and leave the farm. They cannot be held accountable anymore. CSA members – the food consumers – provide capital through their member pledges before each season and they share the risks of poor harvest and mismanagement. So the burden is not carried by the individual farmers (who get into debt in the case of Regionalwert AG). At the same time, members/consumers can no longer hold them accountable by either buying or not buying the food (membership pledges are paid in advance). The high turnover among people working on the farm indicates their freedom and non-accountability vis-à-vis consumers. In addition, as Luzernenhof owns most of the cultivated land based on unconditional loans, there is only an obligation towards third party landowners for a few additionally rented plots of land.

In the cases of both Regionalwert AG and Luzernenhof, there is less burden on the actual farmers and processors (business owners) – at the expense of investors who are willing to take high risks without compensation and at the expense of the consumers/members who pay higher prices for their food. Investors are the "citizen" shareholders in the case of Regionalwert AG, and CSA members, donors and the two secret lenders in the case of Luzernenhof. In both projects and among groups, gains and risks are distributed unequally. Regionalwert AG business owners carry more burden in the form of debt than their employees. In return, they can basically keep potential profits for themselves. At Luzernenhof, analogously, the three farmers who legally own the farm receive the highest salaries. Their actual risks are however modest or non-existent, as the farm does not need to produce in an economically sustainable manner, settle a debt or pay land rents.

Despite similarities, the two cases offer very different pathways. Regionalwert AG aims for a comprehensive accounting system in which risks/return, including ecological and social risk/return, are renumerated and financially compensated. Agency and accountability would increasingly be defined through the market. Luzernenhof is a trust-based project. People devote their time and capital to an idea, rather than to the exchange of goods and services. This allows them to share risks/returns without definite agency and accountability.

Conclusion

With regard to the agri-food system of the Freiburg region, we may argue that the system is currently at the first stage of release. Given the backdrop of an environmentally conscious "green city", people have recognized the problems of the global industrial agri-food system. There is a huge consumer demand for local and organic food. In this chapter, I presented two viable alternatives in the form of a less radical market alternative and a more radical "anti-capitalist" alternative. Both provide different avenues of agency and empowerment as well as risks of exclusion and lack of accountability.

When and how did agents matter for Regionalwert AG and Luzernenhof? Who is empowered through these projects? In both cases, farmers took agency and reorganized farm systems in a way that "empowered" them to produce food locally and organically. As a farmer, Hiss went out to find "citizen" shareholders. Similarly, at Luzernenhof the Widmaiers first had to find volunteers for the farm team, and the farm team then had to attract CSA members/consumers to make pledges. In addition, Luzernenhof found donors and the two lenders who gave them unconditional loans. While farmers hence "got things done", this would not, however, have been possible without "new thinking" among investors and consumers who became willing to accept a higher share of the burden of local and organic food production in the form of low or no capital return and higher food prices. There has been "new thinking" – and, hence, self-determined agency on the side of food producers as well as of consumers.

Does food localization prevent and/or create new prospects for self-determined agency? Financial capital and investments were crucial to both projects. Those people who provided the capital with or without expecting repayment were key when it came to deciding which path to take. In the case of Regionalwert AG, this is obvious as "citizen" shareholders have an institutionalized voice through regular shareholder meetings. As partner companies are accountable to their shareholders, farmers are subject to structural constraints defined by the market. This is different for Luzernenhof. CSA projects leave more self-determined agency to farmers. They do not owe anything to the "investors" and can leave the farm at any time (even during harvest season).

In the case of Luzernenhof, lenders and donors are not made transparent and lack the rights respectively provided by Regionalwert AG. In particular, the two secret lenders support the idea of a non-market-based agri-food system without being repaid and without being supplied with the farm products themselves. CSA members, who through their pledges bear the risks of farming, lack crucial information to actively participate in farm development and hold farmers accountable for (potential) mismanagement. If they do not like the cheese, for instance, they need to have additional resources to buy another type of cheese in the supermarket.

While it is possible to make pledges below the average, and some Luzernenhof members spend less on food compared to what they would pay in the conventional (organic) market, this works only for a limited number of members who are "subsidized" by others. In consequence, CSA membership is generally exclusive to those who can afford the risk of paying twice for their food in the case of failed harvest, or simply different tastes. Likewise, not all Freiburg citizens have the capital to become "citizen" shareholders. So while both projects allow for new agency and empowerment, they risk excluding less affluent people. Indeed, while Luzernenhof and Regionalwert AG provide an opportunity for citizens to participate in food localization, both projects rely on the "new thinking" of people in the region with material capabilities.

Notes

1 Translation from German: "Mir geht es nicht um Gemeinnützigkeit. Die Regionalwert AG ist kein Spendensammler, sondern ein Wirtschaftsunternehmen. Geld ist wichtig."

2 Translation from German: "Ich habe nicht gesagt, Scheißkapitalismus, und habe dann aufgegeben, sondern gelernt, mit seinen Instrumenten zu arbeiten."
3 Translation from German: "Es geht nicht nur darum, dass die Mitglieder uns Geld geben und dafür Milch, Joghurt, Käse bekommen [...] sondern auch darum, dass sie eine Idee unterstützen. Das ist manchmal schwer zu vermitteln."
4 Translation from German: "Theoretisch könntest du bei der Mitgliederversammlung schon Fragen stellen. Aber tatsächlich sitzt du nur da. Du vertraust Johannes. Und du fühlst dich schlecht, weil du keine höheren Beiträge zahlen kannst."

References

Barnett, M. and Duvall, R. (2005). Power in Global Governance. In: M.N. Barnett and R. Duvall (eds.), *Power in Global Governance*. Cambridge: Cambridge University Press, pp. 1–32.

Blum, M. (2010). Rendite, die alle reich macht. *Brandeins* (2). Available at: www.brandeins.de/archiv/2010/marke/rendite-die-alle-reich-macht/ (accessed 1 March 2015).

Boasson, E.L. and Wettestad, J. (2014). Policy Invention and Entrepreneurship: Bankrolling the Burying of Carbon in the EU. *Global Environmental Change*, 29(6), 404–412.

Bognanni, M. (2010). Ein Manager investiert ins Grüne: Serie: Die Bessermacher. *ZEIT online*, 23 August 2010. Available at: www.zeit.de/wirtschaft/unternehmen/2010-08/regionalwert (accessed 25 February 2016).

Buehler, R. and Pucher, J. (2011). Sustainable Transport in Freiburg: Lessons from Germany's Environmental Capital. *International Journal of Sustainable Transportation*, 5(1), 43–70.

Butler, J. (2004). *Precarious Life: The Powers of Mourning and Violence*. London: Verso.

Finnemore, M. and Sikkink, K. (1998). International Norm Dynamics and Political Change. *International Organization*, 52(4), 887–917.

Frey, W. (2011). *Freiburg Green City*. Freiburg i.B.: Herder.

Freytag, T., Gössling, S. and Mössner, S. (2014). Living the Green City: Freiburg's Solarsiedlung between Narratives and Practices of Sustainable Urban Development. *Local Environment: The International Journal of Justice and Sustainability*, 19(6), 644–659.

Grenier, P. (2008). Social Entrepreneurship: Agency in a Globalized World. In: A. Nicholls (ed.), *Social Entrepreneurship. New models of Sustainable Social Change*. Oxford: Oxford University Press, pp. 119–143.

Hardt, M.-X. (2012). Der geteilte Bauernhof: Gemeinschaftliche Landwirtschaft in Seefelden. *Fudder*, 4 May 2012. Available at: http://fudder.de/artikel/2012/05/04/der-geteilte-bauernhof-gemeinschaftliche-landwirtschaft-in-seefelden/ (accessed 25 February 2016).

Hayward, C. and Lukes, S. (2008). Nobody to Shoot? Power, Structure, and Agency: A Dialogue. *Journal of Power*, 1(1), 5–20.

Hodeige, C. (2015). Wie kann der ökologische Landbau vorangebracht werden? *Badische Zeitung*, 15 January 2015. Available at: www.badische-zeitung.de/baden-x7x/wie-kann-der-oekologische-landbau-vorangebracht-werden--99714525.html (accessed 25 February 2016).

Holling, C.S. (ed.) (1978). *Adaptive Environmental Assessment and Management*. Chichester, UK: Wiley.

Jordan, A. and Huitema, D. (2014). Policy Innovation in a Changing Climate: Sources, Patterns and Effects. *Global Environmental Change*, 29, 387–394.

Kronsell, A. (2013). Legitimacy for Climate Policies: Politics and Participation in the Green City of Freiburg. *Local Environment: The International Journal of Justice and Sustainability*, 18(8), 965–982.

Luzernenhof (2016). *Home page*. Available at: www.luzernenhof.de/ (accessed 25 February 2016).

Mair, J., Robinson, J. and Hockerts, K. (eds.) (2006). *Social Entrepreneurship*. London: Palgrave Macmillan.

Mintrom, M. and Norman, P. (2009). Policy Entrepreneurship and Policy Change. *Policy Studies Journal*, 37(4), 649–667.

Moore, M.-L. and Westley, F. (2011). Surmountable Chasms: Networks and Social Innovation for Resilient Systems. *Ecology and Society*, [online] 16(1), 5. Available at: www.ecologyandsociety.org/vol16/iss1/art5/ (accessed 22 August 2018).

Moore, M.-L., Westley, F., Tjornbo, O. and Holroyd, C. (2012). The Loop, the Lens, and the Lesson: Using Resilience Theory to Examine Public Policy and Social Innovation. In: A. Nicholls and A. Murdock (eds.), *Social Innovation*. Basingstoke, UK: Palgrave Macmillan, pp. 89–113.

Partzsch, L. (2017). "Power with" and "Power to" in Environmental Politics and Transitions to Sustainability. *Environmental Politics*, 26(2), 193–211.

Pitkin, H.F. (1985). *Wittgenstein and Justice: On the Significance of Ludwig Wittgenstein for Social and Political Thought*. Berkeley, CA: University of California Press.

Prummer, K. (2011). Früchte des Zorns. *Financial Times Deutschland*, 19 December 2016. Available at: www.regionalwert-ag.de/wp-content/uploads/2014/05/FTD_19Dez2011_Social_Entrepreneur-des_Jahres.pdf (accessed 1 March 2016).

Regionalwert AG (2016). *Home page*. Available at: www.regionalwert-ag.de (accessed 25 February 2016) .

Resilience Alliance (2016). *Home page*. Available at: www.resalliance.org/adaptive-cycle (accessed 25 February 2016).

Saurer, M. (2012). Rendite gibt's in Naturalien. *Badische Zeitung*, 21 February 2012. Available at: www.badische-zeitung.de/kreis-breisgau-hochschwarzwald/rendite-gibt-s-in-naturalien-56048171.html (accessed 25 February 2016).

Schindler, H. (2014). Gründer der Regionalwert AG: Erst ein Buch, dann neue Aktien. *Badische Zeitung*, 12 June 2014. Available at: www.badische-zeitung.de/freiburg/gruender-der-regionalwert-ag-erst-ein-buch-dann-neue-aktien-86136083.html (accessed 1 March 2016).

Shey, J. and Belis, D. (2013). Building a Municipal Food Policy Regime in Minneapolis: Implications for Urban Climate Governance. *Environment and Planning C: Government and Policy*, 31(5), 893–910.

Siebold, H. (2011). Unternehmer vom Kaiserstuhl mit Sozialpreis ausgezeichnet. *Badische Zeitung*, 16 December 2011. Available at: www.badische-zeitung.de/wirtschaft-regional/unternehmer-vom-kaiserstuhl-mit-sozialpreis-ausgezeichnet-53571888.html (accessed 25 February 2016).

Steckmeister, J. (2012). Heute schon an (über-)morgen denken. *Badische Zeitung*, 11 December 2012. Available at: www.badische-zeitung.de/eichstetten/heute-schon-an-ueber-morgen-denken (accessed 1 March 2016).

Stuttgarter Zeitung (2013). Der Luzernenhof im Markgräflerland. *Stuttgarter Zeitung*, 4 July 2013. Available at: www.stuttgarter-zeitung.de/inhalt.print.f5ccb0c2-bc36-43c1-a9e2-11e497d5e92d.presentation.print.v2.html (accessed 25 February 2016).

Wendt, A. (1992). Anarchy is What States Make of It: The Social Construction of Power Politics. *International Organization*, 46(2), 391–425.

Westley, F., Olsson, P., Folke, C., Homer-Dixon, T., Vredenburg, H., Loorbach, D., Thompson J., Nilsson, M., Lambin, E., Sendzimir, J., Banerjee, B., Galaz, V. and Van Der Leeuw, S. (2011). Tipping Toward Sustainability: Emerging Pathways of Transformation. *AMBIO*, 40(7), 762–780.

Westley, F., Patton, M.Q. and Zimmerman, B. (2007). *Getting to Maybe: How the World is Changed*. Toronto: Vintage Canada.

Zehfuss, M. (2007). Subjectivity and Vulnerability: On the War in Iraq. *International Politics*, 44(1), 58–71.

Part II

Local agri-food systems

The ideal model of a short food supply chain (SFSC) develops dense ties and linkages within its wider area. But is this the case for actual SFSCs operating in the Global North? And what are the main orientations of the actors in SFSCs and the crucial factors across different geographical settings that define their attributes and success? The four chapters in this part aim to provide answers to these questions by exploring the dynamics of various types of local agri-food systems and their actual linkages within their localities. The first two chapters focus on independent agri-food micro-enterprises operating in strikingly differentiated settings such as Ontario, Canada and the hinterland of the greater Athens region in Greece. The following two chapters focus on two specific types of SFSCs that bear Geographical Indication (GI): the world renowned Parmigiano-Reggiano cheese in Italy and Feta Thessalia in Greece.

5 The long and the short of it

Motivations and realities for food hub actors in Ontario, Canada

Alison Blay-Palmer, Erin Nelson, Phil Mount and Mike Nagy

Introduction: fostering sustainable food systems through transformation

While it is becoming increasingly obvious that we need to transform the conventional food system before we can attempt to build more sustainable alternatives (McInnes and Mount 2017), the complex and interwoven nature of food systems presents both challenges and opportunities. Food has the capacity to be – at once – local economic development and global investment opportunity; source of good or ill health; and climate change solution and contributor. A transformation towards sustainability requires that we aspire to regenerative ecologies, economic localization and equitable access to healthy food, while encouraging democratic engagement and participation across these systems and throughout society (Feenstra 2002; Blay-Palmer and Koc 2010; Anderson *et al.* 2017).

The question this chapter explores is how to think about sustainable food system transformation in the context of the observation: "A complete transformation of the agriculture and food systems, it might be argued, requires a complete transformation of the society" (Magdoff *et al.* 2000:188). When considering how to do this, it is important to acknowledge that food systems are intimately entwined with their places (Franklin and Marsden 2014; Sonnino *et al.* 2016) and so regional and local considerations are important in developing sustainable food system (SFS) approaches. Accordingly, this chapter reports the findings from our empirical work complemented by two international examples that together highlight factors that can enhance the capacity and operations of local sustainable food hubs and show how they can be more impactful.

There is an increasing acknowledgement that food system challenges are complex and interwoven. We can draw direct lines between food and multiple dysfunctionalities: malnutrition as a lack of food causes stunted and decreased human opportunities through the depressed capacity for physical and cognitive growth of tens of millions of children every year (Thurow 2014; WHO 2018); corporate concentration threatens the seed commons through corporate mergers, e.g. between Bayer and Monsanto (IPES 2017);

and, the reductionist, technological turn to nutritionism and the nanotization of food offer "solutions" that are out of reach for the majority of the world's population and depend on highly intensive resource use (Scrinis and Lyons 2010; Scrinis 2013). While food is necessary for human well-being, it is also now among a suite of investment opportunities in derivatives markets. As this financialization of our food systems has occurred, civil society has not been able to engage as effectively as it might have done given the opaque interactions at the nexus of finance and food. This, in turn, makes clear public discourse about the potential for more sustainable food systems more difficult (Clapp 2014: 811). Food is also linked to climate change as long global supply chains, extensive use of chemical fertilizers and improper food waste disposal are significant contributors to global greenhouse gas emissions. At the same time as food systems contribute to climate change, they are also impacted by it – for example, through increased droughts, heat waves, storms and other severe and unpredictable weather events. Issues such as these merely scratch the surface of the challenges faced. Given the "wide range of actors" (Flynn and Bailey 2014) at multiple scales throughout the food system, the complexities are vast. The connections seem too complicated to unravel and it is difficult to see how to bring about change.

Taking a more optimistic vantage point, food is multi-functional and as such it offers multiple opportunities to address health, ecological, economic and civic engagement challenges (Lyson 2008; Morgan 2010). Food systems are increasingly viewed as levers for change at multiple scales. Short food supply chains, city region food systems and alternative food systems, among other approaches, offer entry points into the analysis, practice and theory of sustainable food systems at a community scale (Mount 2012; Blay-Palmer *et al.* 2018). This literature converges around the need for closer links between producers and consumers to enable more localized food systems that provide healthy, clean food, grown and processed using fair and ecological approaches. At the macro scale, the United Nations' Sustainable Development Goals (SDGs) provide a framework for more sustainable food systems. Using SDG 2, the goal to achieve zero hunger by 2030, it is possible to identify multiple points of intersection with other goals and their targets. Obvious links exist with Goal 6 (Clean Water and Sanitation), Goal 12 (Responsible Production and Consumption) and Goal 13 (Climate Action). Other global scale initiatives also support increasingly sustainable food systems, including the Milan Urban Food Policy Pact with more than 160 city signatories. Frameworks such as the New Urban Agenda have weaker ties but also offer promise (Weitz *et al.* 2014; Marsden and Arce 2017; Blay-Palmer *et al.* 2018). Applying a thematic lens, entry points into fostering sustainable food systems include agroecology as striving for ecologically produced food that, among other things, supports local communities and conserves biodiversity (Gliessman 2013; Pimbert 2017); participatory practices that engage people in more democratic food systems processes (Nelson *et al.* 2016); and food policy efforts (Friedmann 2007; Hinrichs and Lyson 2007; Blay-Palmer 2009).

In this chapter, we take local food hubs as our specific entry point for examining the potential to address food-based dysfunctionality through sustainable local food systems. Using survey and case study research in Ontario, Canada, we unpack questions faced at the community scale and provide insights into impediments to food system transformation at both regional and community scales. The challenges identified through our research point to opportunities for change that are then explored through the lens of multi-scaled innovation policies (Scottish Parliament 2015; Morgan 2016), as well as opportunities for fostering network linkages beyond food to amplify a collective sustainable voice through an expanded relational sustainable communities approach (Marsden *et al.* 2010).

Insights from Ontario

Our research project in the province of Ontario, funded by the Ontario Ministry of Agriculture, Food and Rural Affairs (OMAFRA), sought to understand the existing innovation landscape for hubs focused on sustainable local food. The scope and direction of the research was informed by input from community advisors that included practitioners working in small and medium enterprises (SMEs) and civil society organizations (CSOs), government and universities. The research unfolded in two stages. First, we conducted in-depth interviews to build case studies that describe innovation in food hubs along the food chain for sustainable, local food systems (see http://nourishingontario.ca/ontario-food-hub-case-studies-2015/). This case study work was followed by a province-wide survey of actors who self-identified as being part of a food hub initiative. We defined food hubs as, "actual or virtual spaces that collect and distribute food to processors, retailers, restaurants and/or consumers. Food hubs can also provide space for other activities including food preparation, handling, processing, education and/or training." Ultimately, we developed 22 food hub case studies from more than 60 in-depth interviews, and a total of 187 respondents completed a survey that was sent to more than 1,200 individuals and operations.

The case study sites and associated interview participants were selected using a typology that included an assessment of: organizational structure, role in the food chain, motivation and priority objectives, organizational maturity and stability, scale of operations, commodity focus and the relationship to change or innovation. An attempt was made to ensure that cases represented a range across those categories, and to facilitate comparability the case study reports all include sections on: rationale, background information, stories from up and down the supply chain, impacts and benefits, scale and possibilities for expansion, challenges and limitations and keys to success. Building on previous work, we also created organizational maps for some case study hubs, providing organizations with a snapshot of how they are organized and what their networks look like (Mount and Andrée 2013). The case studies were undertaken by region across the province, in northern, eastern and southern Ontario.

Consistent with results from earlier work (e.g. Knezevic *et al.* 2013), common themes were identified through the case studies with respect to both benefits provided by food hub initiatives and gaps to be filled. The community- based benefits included agglomeration effects through economies of scale that resulted in both efficiencies and cost savings. Support for local businesses was also noted as a benefit in many of the case studies. The distribution of more local and regional food, along with increased access to this food and increased food system knowledge, were also seen as positive outcomes in the case studies. The fostering of locally focused partnerships and relationships was also beneficial.

Several common challenges and limitations were also identified across most of the case studies. The most widely reported challenge was the need for more capital. As we will see below from our survey results, this is a challenge that prevents innovation across the organizational spectrum. A range of linked infrastructure issues was also reported. Many innovators noted an insufficient supply of local and local/organic food as well as processed goods. Transportation and distribution resources were also lacking, including trucks and safe food handling facilities. A dependence on volunteer labour was also cited as a problem that is particularly challenging as it feeds into the overall vulnerability of operations that are, in many cases, seasonal. A final commonly identified challenge was competition from conventional retailers and distributors.

We were also able to identify common keys to success for these innovative operations. A gradual building of the operation underpinned by creativity and persistence was cited as a central approach to sustaining and expanding food hub opportunities over time. Key informants said they also had to engage in work and integrate across the food chain to build opportunities. This included engaging with the local community and building relationships across different sectors.

The results from the case studies were confirmed by the province-wide survey. The survey found that respondents are highly motivated to support sustainable food system goals as reflected in a number of responses, including the values held by participants. More than 80 per cent of respondents indicated that responsible, sustainable production and where the product is grown were a high priority, while over 60 per cent selected profitability and social justice as part of a sustainable local food system. Just under 60 per cent indicated the same for citizen engagement.

There was a strong consensus about what constitutes "local" with more than 75 per cent of respondents choosing "within 100 kilometres", the "county" or "region" categories. Given the size of Ontario (more than 415,000 square miles), and Canada more generally, this definition of local is important. In contrast with our respondents' views, OMAFRA considers local to be within the province. In other cases, (e.g. the 2013 Hellman Mayonnaise campaign) local is within Canada (an expanse of more than 3.85 million square miles). Given the high motivation of food system respondents to support regional food it is important to understand the opportunities and barriers to delivering this type of "local" food to eaters.

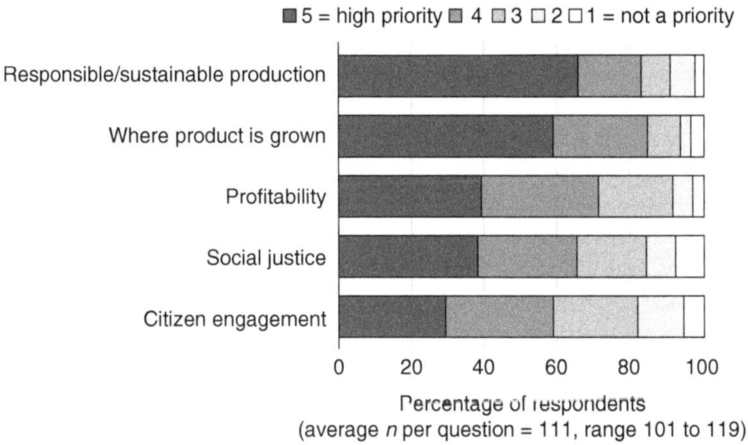

Figure 5.1 Local food values.

In terms of boosting business, 95 per cent of producers surveyed want to increase overall sales, with 56 per cent interested in increasing the amount of land they cultivate. In respect of processors and distributors, 87 per cent reported that they wanted to grow their businesses. To help understand the potential for engaging more with local SMEs, we asked whether the quantity of food purchased from these types of operations was changing. More than half the people surveyed reported steady increases in the amount of food they were purchasing from SMEs, compared with only about 40 per cent who were increasing their purchases from large-scale suppliers. This points to an effort by local sustainable food systems actors to grow their businesses through interactions with SMEs. While there is strong identification with local and sustainable food systems values, and efforts to connect with SMEs are on the rise, barriers to innovation need to be addressed.

The barriers we identified in the survey confirmed those uncovered through previous research in the province (Knezevic *et al.* 2013), and in the case studies discussed earlier. While the median value (the category with the largest number of respondents) for gross annual sales across all respondents is in the C$100,000–C$250,000 category ($n=25$ or 21 per cent), it should be noted that 26 per cent of respondents reported gross annual incomes of C$25,000 or less and this number climbs to 37 per cent when we factor in those with incomes up to C$50,000. These numbers point to financial precariousness for too many operations working in this space. Not surprisingly given the income figures, on the production side, 48 per cent of farmers cited securing capital as their biggest challenge, with nearly 40 per cent also pointing to the need for more access to land, processing infrastructure, consumer education and truck/delivery capacity. The top impediments to expansion across all respondents were: connecting to buyers (more than 75 per cent); financing expansion (more than 60 per cent); increased production (55 per cent); and the challenges of trying to work with large buyers including certification, delivery demands, minimum order requirements, product consistency expectations and timing. These latter challenges suggest that, given the opportunity, there are substantial benefits to working with more SMEs. For processors and distributors, the lack of capital was also the greatest challenge (60 per cent), followed by the need for more processing infrastructure (52 per cent), warehousing space (51 per cent) and staff (49 per cent). Business development support, consumer education and more truck delivery were selected by 37–39 per cent of respondents as needing improvement.

One third of operators indicated the need for more food supply, pointing to opportunities for farmers. By removing barriers, producers could substantially increase production. Farmers projected potential increases of over 135 per cent for meat and poultry, nearly 80 per cent for vegetable production and about 30 per cent for fruit. Farmers also indicated an appetite for growth with 95 per cent wanting to increase their sales. This is particularly hopeful, as a conservative estimate using gross annual incomes produces an estimated *existing* value of local sustainable food systems (LSFS) food hub activity at C$29.3 million.[1] In addition, according to the survey results opportunities for market expansion and growth exist throughout the value chain with 83 per cent indicating that their own storefront provided some to many opportunities for expansion (Figure 5.2).[2] Farmers' markets were a close second with 81 per cent of respondents seeing these as offering opportunities, followed by food co-ops and buying clubs (78 per cent), and then restaurants, caterers and bakeries (73 per cent). One-third or more of respondents saw potential in increasing access to abattoirs, food processing facilities, mobile retail and large grocery stores. With regard to public sector opportunities, 41 per cent of respondents saw long-term care facilities and daycare centres as offering market potential. About one-third of those who completed the survey indicated that colleges and universities (37 per cent), K–12 schools (36 per cent) and hospitals (34 per cent) offered possible sales avenues in the future.

Several benefits are associated with food hubs for producers and communities that point to the direct social, economic and environmental impacts. For producers, the top two existing benefits provided by food hubs were access to new markets and the opportunity to diversify products (see Figure 5.2). Other benefits include educational services, food safety training, increased sustainable production, labelling and certification opportunities, access to demonstration kitchens and liability insurance. These all help producers to be more viable operationally and to reduce their ecological impact. The top two potential benefits identified by farmers that could be provided by food hubs in the future are access to incubator kitchens for product development and the ability to hire more people. Communities also enjoy services related to food hubs (see Figure 5.3). The top services are food donations to local food pantries and food banks, paid employment opportunities for youth

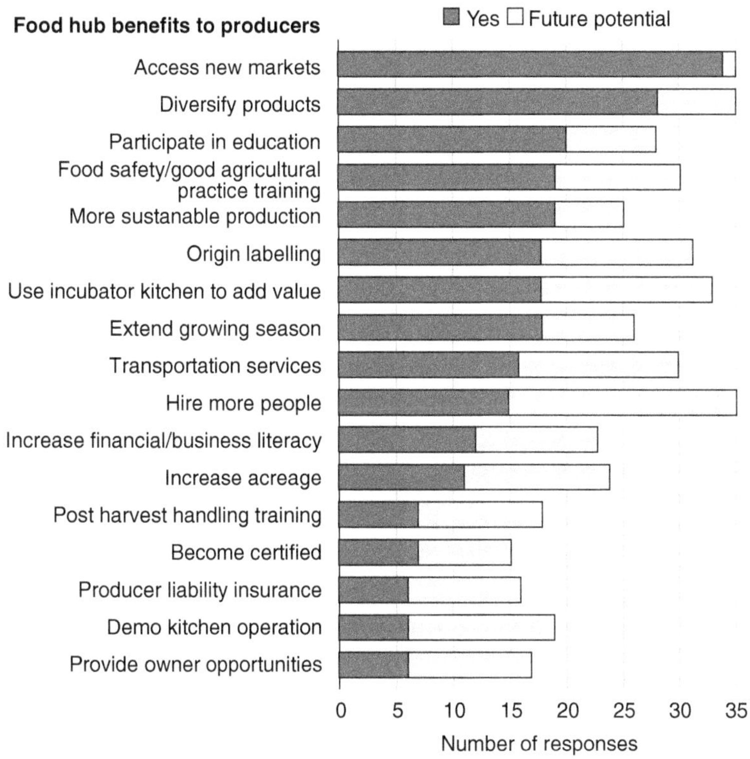

Figure 5.2 Existing and potential food hub benefits for producers.

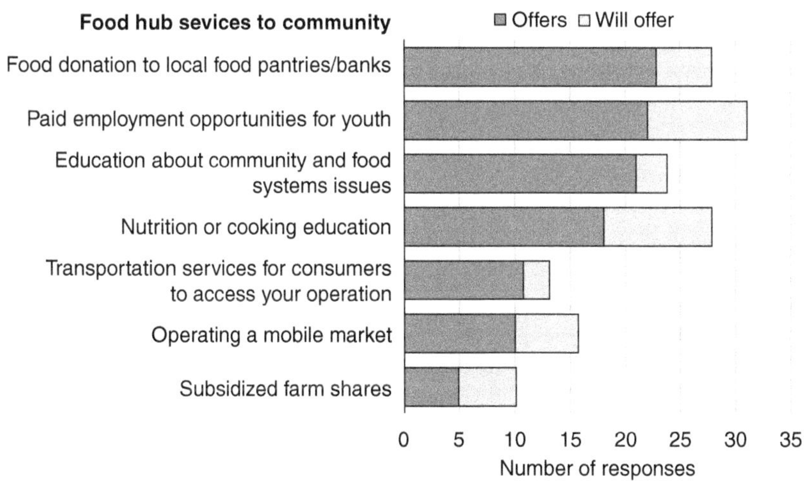

Figure 5.3 Food hub services to communities.

as well as cooking and nutrition education. When we add the operation of a mobile market to this list, these are also where the greatest future potential for food hubs to contribute to communities was identified. Given these opportunities and challenges, it makes sense to consider existing and possible policy options in Ontario and elsewhere to find contexts or examples of how to enable more sustainable food hubs.

Discussion

The results from the Ontario survey and case studies tell the story of community-based food actors from field to fork who aspire to foster food systems that are increasingly sustainable – that is, they are just, embed food in local economies, regenerate the environment and actively engage citizens. Increased food system education, youth employment, food for food banks and pantries and increased local economic activity all contribute to what people increasingly perceive as local sustainable food systems as achieved through food hub-type activity in Ontario. However, while these results confirm that respondents are highly motivated to deliver SFS dividends, there are many challenges to be faced in realizing this vision. Key findings that identify the most pressing gaps include the following needs: greater access to capital; to incentivize and facilitate increased local/sustainable production to boost supply; to facilitate flows of that local supply, e.g. through market channels and infrastructure development; and to enable access to land in support of increased economic activity, production, processing and community benefits that foster *values* chains not simply value chains. As many of these challenges could be addressed through a combination of coordinated institutional and policy responses that build on existing initiatives, it is worth reviewing some of the initiatives currently in place.

Although OMAFRA has been increasingly focused on food exports since 1996, there is an important legacy with respect to supporting a provincial food system. The "Foodland Ontario" brand was developed in the mid-1970s and is recognized by 94 per cent of grocery shoppers in the province (Foodland Ontario 2016). While Foodland Ontario predates contemporary sustainability initiatives, it does signal market opportunities for locally produced and processed food as well as a platform for LSFS if it were adapted to include local or regional branding as well. As a platform for local food, there are separate "food trail" and "food map" initiatives that could be coordinated or integrated with the Foodland brand to better address SFS goals. While promotional efforts for local as part of Foodland do reference farmers' markets, the focus is on large retailers and restaurant chains (Foodland Ontario 2018). On the policy front, in 2013 the Ontario Government passed the Local Food Act, designed to increase access to local food, improve local food literacy, increase local food use by consumers and broader public sector organizations and track uptake on an annual basis. It also includes provisions to establish defined goals for public procurement at some time in the future (OMAFRA

2016). While this sets the context for supporting local food there are many tensions with respect to other elements of food system sustainability. Also, the lack of target setting for public procurement has been disappointing in the years since the act came into effect. In addition to the Local Food Act, the provincial government has also provided more than C$20 million in funding to the Greenbelt Foundation – a charitable organization that provides grants for projects that, among other things, support local food and farming initiatives in a large swath of land in Ontario. While these policies and programmes are or could be used in support of LSFS, more is needed. Accordingly, the next section explores three interventions that could provide directions that would result in accelerated change towards sustainability in Ontario local food systems.

Pulling on the policy threads

The vision expressed in the case studies and survey in Ontario reflects the desire to foster sustainability goals, but also underlines the need for supportive policy. Three promising regional and national policy innovations are discussed below, as ways to support SFS as well as to foster scaling up, as appropriate.

United States Department of Agriculture and Food Hubs

The importance of providing capital for sustainable food systems from the community upwards is demonstrated by initiatives supported by the United States Department of Agriculture (USDA). Over the last ten years or so, USDA has supported hundreds of initiatives in states across the country. For example, the Farmers' Market Promotion Program (FMPP) has provided more than US$58 million to 879 projects since 2008, to fund direct farm to consumer marketing including farmers' markets and agri-tourism initiatives (USDA 2016). This money has been used to provide and improve infrastructure, train farmers and ranchers in direct sell business strategies and to get local healthy food to consumers. In January 2017, an additional US$27 million in funding was announced. These programmes could serve as a guide for support in Ontario to enhance LSFS throughout the province.

Regional innovation policy (RIP): the case of the Basque region

Recently, the Basque region was held up as an entrepreneurial success story by many, including the Organisation for Economic Co-operation and Development (OECD). Given the attention, understanding more about Basque region successes and challenges is valuable. Morgan's research in the Basque region is instructive when considering ways forward for sustainable, local food systems. Grounded in his work on regional innovation systems as well as other research into "smart specialization", Morgan outlines what he calls "collective innovation" that works to coordinate sectoral needs with regional state and the related investment priorities (Morgan 2016: 5). While strategies in the Basque region are focused on science and technology, the overall approach could gain traction for regional SFS work as well.

The regional regeneration process in the Basque region initiative developed in two phases. Policies to support "doing better things" in the 1990s were fashioned to create both new political structures in the form of decision-making mechanisms and operational change through centres of excellence, and a regional coordinating body was established to provide strategic decision-making support (Morgan 2016: 3). Institutionally, cooperative research centres were used to coordinate existing activities and set future research goals. The most recent streamlined strategy emerged from the European Union S3 approach that was designed to support smart specialization within countries through the identification of good practices to inform policy and strategy development and support data gathering and policy-maker training. The strategy is intended to be evidence-based, with an emphasis on developing strong regional foci, based on strengths relevant for national priorities, linking technology with practice, supporting innovation and risk-taking and encouraging stakeholder engagement (European Commission 2012: 8, cited in Morgan 2016: 6). As Morgan notes, some cautions are needed as the strategy is rolled out. The new plan was layered on top of existing infrastructure resulting in an overly complex set of institutions causing connectivity and collaboration challenges. Given the heavy reliance on government for funding, austerity has also threatened the viability of the innovation system. Nevertheless, these programmes provide insights into policy being integrated with private sector innovation, including what helps to support regional regeneration – critically important when we think about how to improve opportunities for LSFS. Developing policy coherence across scales is a key factor, as is the need for evidence to support policy-making – including relevant farmer and/or traditional knowledge – as well as support to encourage a culture of risk-taking. The last point emphasizes the need to think beyond the conventional food system as we work towards sustainability.

Scotland: on becoming "the good food nation"

In 2009, Scotland launched its first national food and drink policy, "Recipe for Success". As a national, pan-industry initiative, it targeted both export and local market opportunities. It was hailed as a success, with food sector improvements between 2007 and 2011 including an increase of 51 per cent for food and drink exports, a 32 per cent increase in Scottish branded food and drink sold in Great Britain, the addition of 150 new local food initiatives, a 50 per cent increase in Protected Geographical Indication (PGI) products, as well as increased engagement in food education and improved diets (Scottish Government 2014: 6). On the local food front, there were nearly twice as many farmers' markets with a 38 per cent increase in sales, and nearly 300 local food events were reported in 2014. In addition, over 164,000 students and 6,500 teachers learned about healthy cooking practices thanks to the "Healthier Scotland Cooking Bus", and one in three primary schools registered with the "Food for Life Catering Mark", thus moving to healthier, ethical, local food. Finally, the "Zero Waste Scotland" initiative is decreasing waste along the food chain.

For Scotland, the 2014 "Becoming a Good Food Nation" programme was an important next step in the process of food system redefinition. It responded to the national government's acknowledgement that, "We need a Scotland that recognizes, at all levels, the strategic importance of food so we can move towards a healthier, resilient and sustainable food system" (Scottish Government 2014: 17). It is an example of a national food initiative targeted at improving individual health (nearly two-thirds of Scottish adults are overweight), diminishing the environmental footprint of the Scottish food system as a remedy for climate change and diversifying the food and drink industry. These goals are consistent with those for Ontario for both the conventional food system and the LSFS.

Following the definition of a "Good Food Nation" and establishing the possibility of creating a Food Commission, the Scottish Government initiated a public consultation process to develop the programme. Public input was sought through online surveys and organizational initiatives and focused on the vision of a "Good Food Nation", its potential impact, the role of a Food Commission and indicators to measure progress. Organizational responses came from across public, private and third sectors including local and national authorities with mandates related to health, environment and sustainability, and from commercial producers, processors and retailers. Respondents supported the broad reach of the vision and appreciated the multiple dividends this vision could help to realize. The need to reform the culture of food in Scotland was widely supported within the context of clearly defined terms of reference, metrics and long-term goals.

In terms of overall context, the consultation analysis set out the complexity of the policy environment including the range of stakeholders with competing interests and priorities and the need for a grounded, achievable proposal that would help to realize success (Scottish Parliament 2015: 2). Key tensions that emerged during the consultation process included reconciling negative environmental impacts and economic growth, and establishing a balance between promoting both local and export market development. Despite these challenges, there was overall support for sustainable economic growth and the need for concerted attention to ensuring food security for everyone, as well as the need for a specific action plan to implement the vision.

Delivery and implementation mechanisms identified through the consultation process included the imperative to involve and coordinate across public, private and third sectors and to engage the full range of policy and programme levers, while using a top down–bottom up approach that is "inclusive, holistic, integrated and bold" (Scottish Parliament 2015: 3). Identifying local good practices and learning from initiatives outside Scotland was also seen to have benefits. There was widespread support for a Food Commission, which was announced in 2015 and includes 15 members with expertise in the environment, retail, seafood, food security, public food, the food industry, access to affordable food, civil society, local food, education, public health and food safety as well as chefs, farmers and youth (Scottish Parliament 2015). The Commission's mandate includes championing and advocating for the place of food in Scotland with respect to overall sustainability including health, environment, the economy and well-being, and moving forward measures that will help the country to become a "Good Food Nation" (Scottish Government 2014: 19). Finally, people responding to the public consultation affirmed that the vision behind the "Good Food Nation" required policy support, a credible action plan supported by adequate funding and political support (Scottish Parliament 2015: 4).

The cases from the Basque region and Scotland highlight the potential for LSFS success in the context of supportive policy and public consultation, and point to the benefits of iteration and coordination between policy scales and institutions, the need to strategically enhance markets and the positive results of more robust focused policy.

Conclusion: moving to sustainable food system change and community integration

The research reported from Ontario confirmed excellent examples of innovation across the local, sustainable food systems chain. Both case study and survey work point to benefits and challenges associated with the emergence of LSFS (Table 5.1).

Table 5.1 makes it clear LSFS are multi-functional and provide levers to address various community and systems challenges that, if addressed, could bring about sustainability change. The benefits include improved economic activity in the form of increased business income, more youth employment and diversified market opportunities. Communities are benefitting through improved food literacy, increased availability of local sustainable food and more local food being delivered to food banks and pantries. The research also points to latent opportunities to grow beyond the estimated C$30 million in annual gross income and add even more to the social fabric and environmental well-being of communities.

All our research points to access to capital as the missing key ingredient for the vast majority of food hub innovators. The need for capital touches on the operational ability to expand, but also to build missing infrastructure that could potentially transfer more benefits to communities. The connection between policy-based support and practical needs is central to achieving lasting results. The Ontario Local Food Act and initiatives through

Table 5.1 Benefits and challenges associated with LSFS in Ontario: findings from case studies and surveys

	Benefits	*Challenges*
Case studies	Increased efficiency and cost savings through economies of scale. Increased availability of local, sustainable food. Improved food access. Enhanced networks and partnerships. Improved food system knowledge.	Dependence on volunteer labour. Competition from conventional retailers and distributors.
Surveys	Desire to increase local sales. Notable economic impact of ~$30 million/year. Steady increase in business with SMEs. Potential production increases of over 135% for meat and poultry, nearly 80% for vegetable production and around 30% for fruit. Opportunities for expansion exist for storefront operations and farmers' markets as well as food co-ops and buying clubs; there are some opportunities across the public sector. For producers, opportunity to diversify their products with potential benefits of access to incubator kitchens for product development and ability to hire more people. Benefits for communities are: food donations to local food pantries and food banks; paid employment opportunities for youth as well as cooking and nutrition education; and mobile markets – these are where the greatest future potential was identified for food hubs to add to communities.	Producers need access to more land and support through better consumer education. For processors and distributors, more warehouse space and more staff are required. Connections to buyers and challenges working with large-scale buyers.
Both case studies and surveys	Support for local businesses. Access to new markets for producers.	Lack of capital with precarious annual gross incomes for up to 26% of respondents. Lack of supply of local and/or organic fresh and processed food. Need improved distribution infrastructure including trucks and safe handling facilities.

OMAFRA demonstrate an opening where support could create more vibrant sustainable local food initiatives. More coherent and joined up policies would enable next steps (MacRae 2011). In Scotland, "Becoming a Good Food Nation" demonstrates how joined up policy and vision can help move the compass increasingly towards LSFS even in the context of export market activity. The regional innovation strategies in the Basque region offer examples of "doing better things" and of models for creating new structures and operational supports (Morgan 2016: 3) by linking policy to action on the ground. In the Canadian context, MacRae (2011) suggests that there are three revenue streams that could be tapped into to support shifts to more joined-up policy and that this initiative could be led by the federal government. The first involves shifting from subsidies for export-focused production to subsidies for SFS. Second, MacRae suggests that if producers moved to low-input production systems that regenerate land and water to create more robust and resilient production systems, there would be less pressure on government for financial assistance due to on-farm crises. And, finally, pivoting the health care system from care to prevention through healthier food systems could also reduce costs (MacRae 2011: 454).

Reaching beyond the local, international initiatives such as the Milan Urban Food Policy Pact and the New Urban Agenda provide points of intersection across multiple scales and offer a way to develop networks internationally that facilitate both recognition of the potential of LSFS as well as knowledge sharing opportunities (Blay-Palmer *et al.* 2018). While LSFS are emergent in Ontario, the research presented in this chapter identifies specific ways forward based on research in the Basque region, Scotland and on the USDA experience, and provides excellent examples of how more joined-up policy supported by targeted funding could help to embed and grow LSFS in Ontario and elsewhere.

Notes

1 Estimates were calculated using midpoints for the income brackets up C$250,000 and the lowest value for each category was used thereafter. These values were multiplied by the number of respondents in each category.

2 Data note: for the question related to opportunities for growth, percentages were calculated based on the number of respondents in each category.

References

Anderson, C., Brady, J. and Levkoe, C. (2017). *Conversations in Food Studies*. Winnipeg: University of Manitoba Press.

Berti, G. and Mulligan, C. (2016). Competitiveness of Small Farms and Innovative Food Supply Chains: The Role of Food Hubs in Creating Sustainable Regional and Local Food Systems. *Sustainability* 8: 616. Doi:10.3390/su8070616.

Blay-Palmer, A. (2009). The Canadian Pioneer: The Genesis of Urban Food Policy in Toronto. *International Planning Studies* 14(4): 401–416.

Blay-Palmer, A. and Koc, M. (2010). Imagining Sustainable Food Systems: The Path to Regenerative Food Systems. In: A. Blay-Palmer (ed.), *Imagining Sustainable Food Systems*. Farnham, UK and Burlington, VT: Ashgate Publishing.

Blay-Palmer, A., Sonnino, R. and Custot, J. (2016). A Food Politics of the Possible? Growing Sustainable Food Systems through Networks of Knowledge. *Agriculture and Human Values* 33(1): 27–43.

Blay-Palmer, A., Santini, G., Dubbeling, M., Renting, H., Taguchi, M. and Giordano, T. (2018). Validating the City Region Food System Approach: Enacting Inclusive, Transformational City Region Food Systems. *Sustainability* 10(5): 1680.

Clancy, K. (2014). A Different Way to Approach Policy Change. *Journal of Agriculture, Food Systems, and Community Development* 4(4): 9–11. Available at: http://dx.doi.org/10.5304/jafscd.2014.044.010.

Clapp, J. (2014). Financialization, Distance and Global Food Politics. *Journal of Peasant Studies* 41(5): 797–814.

European Commission (2012). Guide to research and innovation strategies for smart specialisations. RIS3 Guide, Brussels.

Feagan, R. (2007). The Place of Food: Mapping Out the "Local" in Local Food Systems. *Progress in Human Geography* 31(1): 123–142.

Feenstra, G. (2002). Creating Space for Sustainable Food Systems: Lessons from the Field. *Agriculture and human values* 19(2): 99–106.

Flynn, A. and Bailey, K. (2014). 5 Sustainable Food Supply Chains. In: T. Marsden and A. Morley (eds.), *Sustainable Food Systems: Building a New Paradigm*. London and New York: Routledge, p. 103.

Foodland Ontario (2016). www.ontario.ca/foodland/page/results-consumer-testing.

Foodland Ontario (2018). Foodland Ontario: Finding Local. Available at: www.ontario.ca/foodland/page/finding-local.

Franklin, A. and Marsden, T.K. (2014). (Dis)connected Communities and Sustainable Place-Making. *Local Environment: The International Journal of Justice and Sustainability* 20(8): 940–956. Doi: 10.1080/13549839.2013.879852.

Friedmann, H. (2007). Scaling Up: Bringing Public Institutions and Food Service Corporations into the Project for a Local, Sustainable Food System in Ontario. *Agriculture and Human Values* 24(3): 389–398.

Gliessman, S. (2013). Agroecology: Growing the Roots of Resistance. *Agroecology and Sustainable Food Systems* 37(1): 19–31.

Hinrichs, C. and Lyson, T. (eds.) (2007). *Remaking the North American Food System: Strategies for Sustainability*. Lincoln, NE: University of Nebraska Press.

Holt-Giménez, E. and Shattuck, A. (2011). Food Crises, Food Regimes and Food Movements: Rumblings of Reform or Tides of Transformation? *Journal of Peasant Studies* 38(1): 109–144.

IPES (International Panel of Experts on Sustainable Food Systems) (2017). Too big to feed: Exploring the impacts of mega-mergers, consolidation and concentration of power in the agri-food sector. Available at: www.ipes-food.org/images/Reports/Concentration_FullReport.pdf.

Knezevic. I., Blay-Palmer, A., Landman K. and Nelson, E. (2013). Models and best practices in Ontario communities of food. Available at: http://nourishingontario.ca/reports/.

Levkoe, C. (2011). Towards a Transformative Food Politics. *Local Environment* 16(7): 687–705.

Levkoe, C. (2014). The Food Movement in Canada: A Social Movement Network Perspective. *Journal of Peasant Studies* 41(3): 385–403. Doi: 10.1080/03066150.2014.910766.

Levkoe, C. and Wakefield, S. (2014). Understanding Contemporary Networks of Environmental and Social Change: Complex Assemblages within Canada's "Food Movement". *Environmental Politics* 23(2): 302–320.

Lyson, T.A. (2008). Agriculture of the Middle: Lessons Learned from Civic Agriculture. In: T.A. Lyson, G.W. Stevenson and R. Welsh (eds.), *Food and the Mid-Level Farm: Renewing an Agriculture of the Middle*. Cambridge, MA and London: MIT Press, pp. 165–178.

McInnes, A. and Mount, P. (2017). Actualizing Sustainable Food Systems. In: C. Anderson, J. Brady and C. Levkoe (eds.), Conversations in Food Studies. Winnipeg: University of Manitoba Press, pp. 332–347.

MacRae, Rod. (2011). A Joined-Up Food Policy for Canada. *Journal of Hunger & Environmental Nutrition* 6(4): 424–457.

Magdoff, F., Bellamy Foster, J. and Buttel, F.H. (eds.) (2000). *Hungry for Profit: The Agribusiness Threat to Farmers, Food, and the Environment*. New York: Monthly Review Press.

Marsden, T. and Arce, A. (2017). The Social Construction of International Food: A New Research Agenda. In: R. Munton (ed.), *The Rural*. London: Routledge, pp. 87–106.

Marsden, T., Franklin, A., Newton, J. and Middleton, J. (2010). Sustainability in Practice: Situated Learning and Knowledge for the Evolving Eco-Economy. *Town Planning Review* 81(5): 541–562.

Morgan, K. (2010). Local and Green, Global and Fair: The Ethical Foodscape and the Politics of Care. *Environment and Planning A* 42(8): 1852–1867.

Morgan, K. (2016). Collective Entrepreneurship: The Basque Model of Innovation. *European Planning Studies* 24(8): 1544–1560. Doi: 10.1080/09654313.2016.1151483.

Morgan, K. and Sonnino, R. (2008). *The School Food Revolution*. London: Earthscan.

Mount, P. (2012). Growing Local Food: Scale and Local Food Systems Governance. *Agriculture and Human Values* 29(1): 107–121.

Mount, P. and Andrée, P. (2013). Visualizing Community-Based Food Projects in Ontario. *Local Environment* 18(5): 578–591.

Mount, P., Hazen, S., Holmes, S., Fraser, E., Winson, A., Knezevic, I., Nelson, E., Ohberg, L., Andrée, P. and Landman, K. (2013). Barriers to the Local Food Movement: Ontario's Community Food Projects and the Capacity for Convergence. *Local Environment* 18(5): 592–605.

Nelson, E., Tovar, L.G., Guegen, E., Humphries, S., Landman, K. and Rindermann, R. (2016). Participatory Guarantee System and the Re-imagining of Mexico's Organic Sector. *Agriculture and Human Values* 33(2): 373–388.

OMAFRA (2016). 2015/2016 Ontario Local Food Report. Available at: www.omafra.gov.on.ca/english/about/local_food_rpt16.htm.

Pimbert, M. (ed.) (2017). *Food Sovereignty, Agroecology and Biocultural Diversity: Constructing and Contesting Knowledge*. London: Routledge.

Scottish Government (2014). Becoming a good food nation. Available at: https://beta.gov.scot/binaries/content/documents/govscot/publications/consultation-paper/2014/06/recipe-success-scotlands-national-food-drink-policy-becoming-good-food/documents/00453219-pdf/00453219-pdf/govscot:document/.

Scottish Parliament (2015). *Food (Scotland) Act 2015*. Available at: www.legislation.gov.uk/asp/2015/1/pdfs/asp_20150001_en.pdf.

Scrinis, G. (2013). *Nutritionism: The Science and Politics of Dietary Advice*. New York: Columbia University Press.

Scrinis, G. and Lyons, K. (2010). Nanotechnology and the Techno-Corporate Agro-Food Paradigm. In: G. Lawrence, K. Lyons and T. Wallington (eds.), *Food Security, Nutrition and Sustainability*. London and Sterling, VA: Earthscan.

Sonnino, R., Marsden, T. and Moragues-Faus, A. (2016). Relationalities and Convergences in Food Security Narratives: Towards a Place-Based Approach. *Transactions of the Institute of British Geographers* 41: 477–489. Doi: 10.1111/tran.12137.

Thurow, R. (2014). 1,000 days: The period that decides the health and wealth of the world. *The Atlantic*, 1 May 2014.

USDA (2016). Farmers' Market Promotion Program: 2016 Report. Available at: www.ams.usda.gov/sites/default/files/media/FMPP2016Report.pdf.

USDA (2017). USDA Announces $27 Million in Grants to Support the Local Food Sector, *USDA Agricultural Marketing Service*, news release, 11 January 2017. Available at: www.ams.usda.gov/press-release/usda-announces-27-million-grants available-support-local-food-sector.

Weitz, N., Nilsson, M. and Davis, M. (2014). A Nexus Approach to the Post-2015 Agenda: Formulating Integrated Water, Energy, and Food SDGs. *SAI Review of International Affairs* 34(2): 37–50.

WHO (2018). Stunting in a nutshell. Available at: www.who.int/nutrition/healthygrowthproj_stunted_videos/en/.

6 "New" micro agri-food initiatives in crisis-hit Greece and beyond

A promising alternative or business as usual?

Sophia Skordili

Introduction

Greece has a long-standing specialization in agri-food. Moreover, the sector had maintained its dispersed and somewhat balanced structure until quite recently. However, during the last 30 years or so enduring structural characteristics have been changing fast. Radical changes in the wider business environment have led the fast transition from a small scale and regionalized structure to a corporate-led and mostly urbanized arrangement. These major transformations can be described in two successive shifts.

The first can be placed at the end of the 1980s, with the prospect of the Single European Market (SEM). During the period 1985–1992, the country became a target for expansion for several leading multinational corporations (MNCs) in agri-food processing and retailing. Since then, a small group of multinationals along with a few large domestic competitors are increasing their share in every agri-food product and geographical market in the country. Their easy and uninterrupted growth continued for a period of 25 years, driven by a stable increase in disposable incomes and consumption and the generous support offered by national and EU regional and industrial policies (Skordili, 2013a).

The second shift has been apparent during the persistent crisis that began in 2010 and which has left a heavy footprint in every aspect of social and economic life. The dramatic drop in household disposable incomes by 27.5 per cent (between 2007–2015) has resulted in more than a 30 per cent shrinkage in the food market and changes in consumption patterns (Haralambakis, 2017). In this context, the already highly unequal allocation of markets and profits among different actors in supply chains has been further widened in favour of the bigger competitors. The deep recession has brought the various malfunctions associated with the corporate-led agri-food system into sharper focus. It has become evident that several leading corporations have followed a path of fragile and largely unsustainable growth with long-term damaging effects for people and places.

Next to the powerful leading firms, thousands of weaker actors in the supply chains have gradually been forced to exit the market or struggle to survive in a hostile business environment (INEMY, 2016). However, in the midst of the crisis, the micro agri-food sector has shown divergent trends. Alongside continual market exits, new micro food processors and independent food retailers are emerging all over the country. It is widely believed that a big advantage these new ventures have is that they can act immediately as an antidote to the mounting socio-economic problems associated with corporate-led food production and consumption. They are praised as valuable sources of employment and generators of local development.

This over-optimistic view conflates what is desired and what is real and can be misleading. It stems from a superficial approach that generalizes from a small number of success stories in isolation from the wider business environment. In this case, there are a lot of similarities with the debate on counter-urbanization triggered by unemployment. After several years of high expectations it is now well documented that the original estimations of the "back-to-the-land" trend were grossly overrated (Demetra, 2012; Gkartzios, 2013; Anthopoulou *et al.*, 2017).

There is no doubt that there is much room for micro agri-food initiatives, to provide a concrete response to the widespread concerns about the profound dysfunctionalities of the conventional agri-food system. Yet, there is an urgent need to construct a sharper notion of their actual characteristics and development potential. The aim of the chapter is to shed light on the profile and prospects of the new micro agri-food firm development by exploring two interrelated questions: (i) What is their real capacity, specialization and spatial pattern?; and (ii) How do they actually deliver in terms of employment creation and sustainable local development objectives?

Micro agri-food firms have always been present in large numbers in Greece but they have not escaped criticism (Labrianidis *et al.*, 2004). Small scale and local ownership and control, per se, do not automatically lead to more inclusive, equitable and fair development (Born and Purcell, 2006). The question is whether these "new" ventures bring with them critical new assets that make them qualitatively different from the typical traditional small firm. More specifically, they should espouse certain qualities relating to employment creation, sustainable local development, food accessibility and fairness in order to have a real impact to people and places (Pirog *et al.*, 2014). Sustainability (in its broader sense) and equity are the overarching values of any Good Food chain (Biggeri and Ferrannini, 2014).

This chapter adopts the supply chain approach and focuses on new micro (up to ten employees) agri-food processors and independent food retailers. It is based on an extensive review of analytical and empirical material and the results of a recently conducted research project.[1]

The remainder of the chapter is structured into four parts. The next part consists of a critical review of the impact of small agri-food chains on two crucial areas for crisis-hit Greece: employment creation and sustainable local development. The next two parts are based on the findings of a two-stage research project. The third part presents the results of desk-top research on the capacity, locational pattern and broad profile of the new micro agri-food ventures. The fourth part discusses their actual impact on employment and local development making use of fieldwork material. The final part synthesises the findings and discussion on the development potential of new micro agri-food firms during the crisis and, even more significantly, in the aftermath of eight years of severe austerity and unprecedented economic turmoil.

The impact of local agri-food initiatives on employment and local development

The continuing financial, economic and political turmoil accentuates already existing failures of the prevailing agri-food system, thus creating an urgent need to identify innovative, effective and legitimate responses (Morgan *et al.*, 2006; Blay-Palmer, 2008; Clapp, 2012). During the last two decades, myriads of micro agri-food initiatives have emerged all over Europe and North America. They are found under several contested terms, such as *alternative food networks*, *short food supply chains*, *localized agri-food systems*, *good food chains*, *sustainable food*, *quality food*, *green food*, etc., which relate, not just to particular business practices, but also to value judgements about how the food sector should operate. Notwithstanding their profound differences, they share an opposition to specific failures and harmful effects of the conventional agri-food system. In short, they are different or even countercultural to corporate-led food supply chains (Tregear, 2011). They share a number of common values around food although with notable variations and intensity.

During the first stages of new firm development there was a remarkable, too often exclusive, interest in environmental protection and quality food. This focus has been totally justifiable given the serious environmental and nutritional concerns provoked by the largely unregulated activities of the conventional agri-food system (Lang and Heasman, 2004; Carolan, 2013). However, it soon became evident that these defining values should be extended to include the equally important and rather overlooked social and economic aspects of sustainability. In the context of increased social inequality and restricted household budgets it is crucial that food chains are able provide a valuable source of income and employment to weak actors in supply chains, utilize fair labour practices, boost local economies and supply the market with good quality food at affordable prices. In fact, the overarching question should be: Can micro agri-food initiatives generate economic gains for producers, consumers and local communities, while at the same time satisfying ethical criteria? This is a challenging question with delicate trade-offs and no easy answers.

The hierarchy and relative significance of these purposes varies significantly across different regions (Ilbery and Maye, 2005). The variations can largely be explained by substantial geographical disparities in the organizational structures of food supply chains, consumer perceptions, cultural/gastronomic traditions and institutional and policy support frameworks (Renting *et al.*, 2003).

In north-west Europe, the issues of environmental protection, nutritional value, food safety and provenance are set high on the agenda. At present, however, social and economic goals have crucial significance for the fragile economies of southern and eastern Europe. Certainly, this is the case in crisis-hit Greece. The pressing issues of massive unemployment (around 23 per cent) and persistent recession put the goals of employment creation and sustainable local development high on the agenda. Thus, micro agri-food initiatives need to consider these criteria if they are to play a pivotal role in assisting weak actors in supply chains and leading the country out of the crisis. An assessment of the direct and indirect impact of micro agri-food chains on employment and local economies is, however, a challenging task.

A limited number of studies have attempted to quantify the effects of micro agri-food firms on local economies by making use of tools and methodologies developed in macroeconomics (Otto and Varner, 2005). However, quantitative approaches have been criticized on several grounds. The first focuses on shortages of adequate and reliable data at the micro level, for local areas and small businesses alike. Furthermore, such approaches are charged that they adopt a partial and selective view, and are unable to grasp novel forms of sustainable development that cannot easily be quantified. Abstract qualities can be equally, or even more, important than strictly economic goals that dominate quantitative analysis.

The aforementioned problems constrain the use of quantitative methodologies. Hence, the bulk of research takes the form of localized case studies that make use of qualitative methodologies. Qualitative assessments are valuable sources of information about localized agri-food ventures in different settings. However, they have two key limitations: (i) their results cannot be generalized and do not bare regional comparisons since they refer to particular regional settings and, usually, particular types of micro agri-food initiatives; and (ii) they are mostly based on the perceptions and experiences of the people who take part in the research and can be subjective (Mundler and Laughrea, 2016).

Employment creation

The direct impact of new micro agri-food chains on employment creation is not numerically strong. By definition, micro firms provide modest employment opportunities. Too often, new agri-food firms are a source of formal employment only for their owners. It should also be noted that, due to their familiarity with food, a substantially higher percentage of owners are women; hence these firms' contribution to tackling female unemployment is crucial.

However, new agri-food firms can have a significant indirect impact on safeguarding employment in their wider areas. Micro initiatives around food provide access to the products of vulnerable local producers who have been excluded from conventional food systems or who work in severe price-squeeze conditions (Galli and Brunori, 2013). Hence, the creation of coherent agri-food chains sustains local employment and helps counter the abandonment of agriculture and out-migration.

A sole interest in statistical data can lead to misunderstandings or false conclusions if there is not clear reference to the abstract qualities of labour. Some evidence stresses the positive effects of new agri-food firms in increasing job satisfaction, professional fulfillment and the social recognition of owners (Wiskerke, 2009). Several authors remark that owners have relatively high educational qualifications and their involvement with a new business gives them the opportunity to apply their skills and knowledge (Mundler and Laughrea, 2016). Also, in several cases the start of the business has been cited as a good cause, marking a change in career or the owner's greener lifestyle (Herslund, 2011).

Local development

Mounting research has shown that new agri-food chains deliver a range of economic benefits in local areas, although to varying degrees.

Several studies have shown that new micro agri-food processors guarantee more sustainable use of recourses than conventional food chains (Brunori *et al.*, 2016). They can lead to revitalization of local techniques and local products, making use of local know-how and traditional agriculture and production practices. Hence, places are gradually regaining their defining role in constructing product quality, and local populations can enhance their sense of belonging and local identity through food consumption (Wiskerke, 2009). Finally, too often it is taken for granted that shorter food miles not only contribute to reductions in CO_2 emissions but deliver more nutritious food with less packaging needs (Kneafsey *et al.*, 2013).

In addition, micro agri-food chains contribute to the creation of more solid and resilient local economies. They can enhance the coherence of local production systems, enabling synergies of agri-food with other local/regional activities. The establishment of long-term relations of trust between different actors in the supply chains reduces economic uncertainties and can provide small producers with alternative sources of funding (Sage, 2002). The small number of intermediaries, or even the opportunity for direct sales, provides producers with premium prices when compared to conventional retail and wholesale channels (Galli and Brunori, 2013). The deep relationship between the local food and tourism sectors is well documented (Seems, 2010). These synergies can contribute to the creation of a more diverse economic base, with several types of ownership and firm size, thus enhancing the resilience of the local economic base to external shocks (Sage, 2014).

At the same time, a significant stream of research shows that micro agri-food chains' contribution to sustainable local development can be contested in several respects. The main areas of critique focus on the unsustainable use of resources and weak social embeddedness. Resistance to external forces can lead to the dangerous path of defensive localization (Heinrichs, 2003). In several instances local production has led to environmental pollution and ecological degradation. Studies document that local production is not always the most sustainable option, particularly for products that are not suited to local soil and climate conditions – e.g. tomato production in the UK and southern Spain (Heinrichs, 2003; Born and Purcell, 2006; Brunori *et al.*, 2016). Also, proximity does not automatically lead to trust and community building. Studies from several local settings have shown that interpersonal ties and reciprocity may well coexist with relations of power, conflict and personal gain (Ilbery and Maye, 2005; Allen, 2010).

The capacity and profile of the new micro agri-food chains: desk-top research

As the economic crisis unfolds, the already high concentration ratio within the two key pillars of the domestic agri-food system, food processing and retailing, has deepened further. Recent data shows that the ten leading firms contribute more than 32 per cent to the aggregate sales of the top 100 agri-food processors (Hellenic Manufacturing, 2015). Likewise, the market share of the top three grocery retail chains is remarkable by southern European standards, with 60 per cent of the domestic food market (INEMY, 2016).

The deep recession has brought various malfunctions associated with the corporate-led agri-food system into sharper focus. It has become evident that emblematic leading corporations are leaving their traditional domestic suppliers in favour of cheaper, imported raw and semi-processed products, or even final food products. This trend has led to a spectacular loss of national sufficiency, even for several Mediterranean fruit and vegetables, and the relocation of agri-food processors closer to Athens and Thessaloniki metropolitan conurbations (Skordili, 2013a). The bypassing of rural areas by agri-food corporations has led to the gradual eradication of long-standing local agri-food complexes and the loss of market access for thousands of small farmers. In the sphere of consumption, major retailers have been accused of greed and unethical pricing practices, as supermarket food is relatively expensive in austerity ridden Greece compared to other European countries (Skordili, 2013b).

Hence, there is an urgent need for small-scale localized initiatives to challenge the prevailing agri-food system. There are indications that the contribution of micro firms to employment creation is crucial. More specifically, during 2012–2013 employment in micro firms increased by 8.4 per cent while in big firms (more than 250 employees) it decreased by 1.2 per cent. It is also revealing that, during the same period, micro firms faced a considerable decrease in sales (1.9 per cent) while the respective decrease for big firms was marginal, at only 0.2 per cent (IOBE, 2016).

However, the experience of traditional agri-food small firms has not always been good. They are often charged with producing low quality products as well as intensive agriculture practices, low paid jobs and flourishing informal market transactions. Furthermore, their owners have very low educational qualifications and poor management skills (Labrianidis *et al*., 2004).

A thorough search of various secondary sources took place early in 2015, in order to construct a database of the new micro agri-food processing and food retailing firms operating all over the country. The eligible firms were selected using a range of criteria: they must have started their operations, or been radically reorganized, during the crisis (i.e. after 1 January 2010); they also must have strong associations, either in product qualities or in business organization, with a certain locality. The micro size of the firms was not an issue since the vast majority of the start-ups fulfilled this criterion.

An interesting finding was revealed at the initial stage of research: no more than 150 eligible processing and retailing firms were in operation in mid-2016. Hence, the birth rate of the new firms is considerably lower than is widely believed. In addition, a substantial number of firms were forced to withdraw from the market at the initial stage of development. Lack of access to the banking system, as well as long delays in paying installments in respect of investment projects financed under the EU and state programmes, could have a critical effect on a firm's survival. In several instances, the failure could mostly be ascribed to poor decisions made by the owners. The most notable example is the rapid closure of several start-ups in online food sales because they had underestimated the difficulties, and the logistics, involved in organizing sophisticated supply chains given their small scale and the fragility of fresh food (Bodini, 2014).

Desk-top research was a valuable source of information in the areas of business organization, product specialization and the location pattern of the new firms.

Several firms have internalized several stages of the supply chain despite their small size. They are active in agricultural production and processing, or processing and sales, while a small number are active in all three sectors of economic activity.

There is a remarkable shift in specialization to higher value products. In food processing there is a clear trend in revitalizing traditional local products (wineries, spirits, sauces, cheese and dairy, dried meat), as well as in processing novel products for the domestic market and upmarket food. In retailing, there is a massive influx of speciality local food outlets.

Some authors argue that localized agri-food chains are a product rather than a driver of socio-economic development (Tregear, 2011). The locational distribution of the 150 firms seems to confirm this pattern. There is an intense concentration in the hinterland of Athens and Thessaloniki metropolitan regions. Also, they have a strong presence in areas that combine tourist attraction with significant agri-food specialization, mainly Crete and several other islands, as well as in the southern Peloponnese. Hence, the traditional over-presentation of agri-food processors in certain rural areas of the country seems to be weakening in favour of major urban regions. As a result, the acute regional disparities in the country are expected to widen.

Fieldwork results: the impact of new firms on employment and local development

The outcome of the desk-top research was valuable in enabling a sample to be selected and in providing the geographical focus of the fieldwork research. In total, 30 face-to-face interviews were conducted with the owners of firms registered in the database and located within a two-hour time zone from the centre of Athens metropolitan area. The fieldwork results, however, do not provide a representative picture of the impact of new micro firms. Due to the eligibility criteria for firms and their proximity to the capital city, the sample firms are expected to do better than the national average. A fieldwork survey aimed to explore whether the new initiatives are substantially different, compared to typical traditional small firms, in crucial aspects such as their ability to attract highly qualified labour, launch good quality value-added products, create linkages with local economies and co-operate and form synergies with other firms.

Employment creation

The professional backgrounds of the interviewees indicated that the crisis has triggered three new sources of business owners. First, offspring who had initially chosen a different career path were forced, due to unemployment or instability caused by the crisis, to be involved in the family business later in life. Second, businessmen/women operating in declining sectors (garments, construction, etc.) identified opportunities in the agri-food sector and started a new business. Third, those recently unemployed in several sectors of the service economy (advertising and marketing executives, journalists, architects, teachers, etc.) had started a small agri-food business.

The research shows that low barriers to entry is not the sole factor explaining the incidence of the new ventures. All the business owners that took part in the interviews shared a genuine interest in food and gastronomy. They were curious to discover new products and novel tastes and willing to experiment with new recipes and unknown local ingredients.

The majority of the entrepreneurs that took part in the survey have high educational qualifications. They are holders of university degrees, often at postgraduate level. Also, they are fluent in English and computer skills. However, a substantial number of them have worked only as employees, hence they lack basic entrepreneurial skills such as decision making, management and business planning. Also, the majority of them had only vague knowledge regarding primary production and processing.

The use of additional formal employment was limited. Usually, the owner works long hours with the help, often unpaid, of close family members. They all seem to agree with the view that "*given the instabilities of the wider environment it is too risky to hire employees*" (Interviewee 3).

The inability of firms to integrate in the local economy and society was obvious in several aspects. A large proportion of the entrepreneurs did not have any previous linkage with the wider area in which the processing firm was located. In fact, a not insignificant number of them were Athenians. The decisive factor in starting a business in the area was proximity to the country's main market and use of the advanced transport infrastructure around the capital city. What is more interesting is the fact that they did not move their permanent residence from Athens. They operated their businesses from their Athens offices and visit the processing firms occasionally.

Despite their small size, they were unwilling to cooperate and form synergies with other firms even in respect of basic tasks, such as the common shipment of goods with neighbouring firms to the same retailer. Also, entrepreneurs remained distant from the local community and expressed many complaints about local authorities' lack of support for local firms.

Spatial configuration of supply chains

It was expected, due to the specific eligibility criteria, that the firms interviewed would have developed close linkages and established roots in the local area. However, this was hardly the case at all. In fact, the degree of integration in local areas varies significantly among the firms. To a great extent it depends on the product development strategy used. In search for profit away from highly competitive, saturated and low paid markets, firms adopt various product differentiation strategies. In short, we can identify three distinctive product development strategies: targeting premium and luxury markets; orientation to novel product markets; and slight differentiation in the composition and packaging of existing products. Each strategy entails varying degrees of linkage and embeddedness in the local area. Firms in the two first categories have weak attachments to the local area, while firms in the third category show the closest ties.

Targeting premium and luxury markets

A growing number of new agri-food processing firms have launched premium and luxury products with higher standards in composition, packaging and marketing. There are notable examples in several well-known Greek specialities, mostly in olive oil, with speciality wines, honey and vinegars to follow.

These firms are targeting the upmarket niches of demanding export markets, hence the crucial issue is the construction of an interesting image of the product in order to gain value in the context of symbolic consumption. These firms are mostly initiated by people with previous experience in creative professions in the service sector, who make extensive use of their skills in sophisticated packaging and marketing techniques. The production process is put in second place, or even entirely outsourced on a subcontractual basis. According to one interviewee, "*this olive oil is made from books not from olive trees*" (Interviewee 8).

Production is organized along the lines of extended short food supply chains. The products reach the consumer embedded with value-laden information – too often, too much information – on their quality and production processes. Hence, there is a two-way relationship between business and place. Assets of locality are a strong element of competitiveness and value for firms. In parallel, while the material ties of the firm with the place are very weak, localities take advantage of the reputation they gain as places of origin of premium products.

Launch of novel products

Quite recently a number of novel food products have been launched on the domestic food market by new micro processing firms. Among them, craft beers, edible snails, various types of mushrooms, dips and jams made of various vegetables, etc., have been welcomed by consumers.

The production process for these products is usually an imitation of production and marketing practices already applied in several European countries, notably Italy, France and the Netherlands. Hence linkages with the local area are minimal. In many cases, firms make use of foreign experts, e.g. brewers, and develop close ties with external suppliers and research institutes. However, as the production process goes on, several firms attempt to enhance the local character of the products (e.g. craft beers are infused with new blends of citrus or honey).

One interesting point is the diversification of these firms to consulting services. The first movers in several novel sectors of the Greek market, such as edible snails cultivation and standardisation and truffle hunting and processing, provide training seminars and technical assistance to people who are interested in starting a new business in these sectors.

The great danger for these firms is falling victim of their own success. Several small firms have created new food markets such as craft beers and truffles that have been very successful. Large firms then enter these markets with similar products in order to benefit from the increased demand. As they are better known, have established and extensive market networks, more capital and better marketing, they soon become market leaders at the expense of the first movers – small firms.

Slight differentiation in composition and packaging of existing products

Many new agri-food processing firms stick to the typical product range of small agri-food firms: various types of pasta, jams, spoon sweets (preserves), cookies, and so on. However, their products are substantially differentiated from the well-known and heavily advertised brands of big firms, as well as from the traditional SME products found in the lower part of the market. They produce more sophisticated and value-added products with notable changes in composition (low calorie, high fresh vegetable content, gluten free) and packaging. A significant group of these firms specialize in revitalized, long-forgotten, traditional local products such as carob rusks, local varieties of cheese and cured meat specialities and traditional spirits.

These firms have developed closer linkages with local areas than any other group. In several instances, they are making use of raw agricultural input from the wider area. In this respect, family businesses adopt a unique intergenerational division of labour: the different generations within the owner's family are responsible for different stages of production. As one interviewee commented: "*my parents* [experienced farmers] *run the farm that provides us with the input, my brother-in-law* [industrial designer] *has responsibility for the packaging and logo while I* [MA in Cultural Studies] *am in charge of the sales*" (Interviewee 16).

It should be noted that a not insignificant number of these firms were making use of raw input transported from distant areas of the country. In response to interview questions, entrepreneurs declared that the market asks for *Greek* not *local* food.

The products of these firms combine good quality with affordable prices. They are targeting conscious, middle-class consumers: "*we pay special attention to giving a frugal and no thrills content to our products in order to attract the crisis-hit consumers*" (Interviewee 20). Given the severe recession and increasing market concentration by major supermarket chains it is not surprising that access to the market is the Achilles' heel for the majority of firms. New entrepreneurs were very active in establishing several market channels in order to minimize the risk. More established conventional channels, such as small and medium supermarket chains, coexist with newer and more alternative ones, e.g. direct sales, specialized retailers, local restaurants, etc. Wholesalers are still very active in the micro agri-food sector.

Discussion: prospects and limitations for good food micro chains

The much lower than anticipated number of incidents of micro agri-food start-ups and the many short-lived attempts are strong indications that the public debate on this issue is overly misleading and superficial.

However, the research findings have revealed some really promising signs. In several respects, the firms that took part in the survey are genuinely new compared to the typical traditional agri-food small firm. They provide employment, if limited, and a space for responsible micro entrepreneurship for highly qualified men and women. They contribute to an improvement in quality and to the image of Greek food and gastronomy. In several instances, they have led the regeneration of long-forgotten products and production practices and have developed linkages with local economies. In quite a few respects, new business owners have successfully applied their skills and creativity to adapt to, and transcend, the difficulties posed by the hostile business environment. They have deal with their lack of knowledge in primary production by applying a distinctive intergenerational division of labour within the firm. They add value to their products using low-cost/low-tech differentiation, such as innovative packaging. They increase their revenues with diversification to consulting services.

However, this is not the whole picture. In several less visible but equally important areas, new ventures share the well-known setbacks of typical traditional small firms operating in the domestic agri-food sector. The highly-qualified owners, too often, lack basic business skills and make poor business decisions. With a few notable exceptions, the fieldwork survey revealed weak ties between the firms and local economies and societies. Despite their small size and scale, they were rather unable to form synergies and undertake joint actions. Firms operate, at large, as independent isolated entities and this makes them extremely vulnerable and fragile in such a competitive business environment.

There is no doubt that the persistent deep recession has magnified already existing problems and brought many new complications to the business environment. The lack of access to finance, the predominance of the price criterion on competition and the instabilities of the wider business environment were named as the main causes of failure by business owners. A deeper look reveals two additional areas of concern. First, the crisis hinders business development. Business owners feel too insecure to realize their plans. A notable example is the majority of independent speciality food retailers who do not have the economic means nor the drive to diversify to cooking seminars, new recipe testing, promotion of food from specific localities, etc. Second, there is an obvious retreat of the local provenance of food in favour of domestic products, due to increased market nationalism triggered by the crisis.

There is no doubt that localized micro agri-food systems should play a pivotal role in leading the way out of the persistent deep recession. There is much room for them to provide concrete responses to the widespread concerns of the conventional agri-food system based on the principles of cost-efficiency and negative externalities. Agri-food localization should lead to a food system that is better aligned with the goals of inclusive development, equality, social justice and ecological integrity. However, its long-term viability and success is questionable unless it is supported by territorial food policies suitable at small scale.

Note

1 Desk-top and fieldwork research was carried out in the framework of the THALES research project, "Human capabilities as a basic component of development dynamics and differentiation of the regions: the case of Greece", coordinated by the University of Macedonia (Professor Lois Labrianidis) and co-financed by the

EU (European Social Fund) and the Hellenic Ministry of Education. The author was the coordinator of the Harokopio University team which was responsible for the agri-food sector.

References

Allen, P. (2010). Realizing Justice in Local Food Systems. *Cambridge Journal of Regions, Economy and Society*, 3: 295–308.

Anthopoulou, Th., Kaberis, N. and Petrou, M. (2017). Aspects and Experiences of Crisis in Rural Greece. Narratives of Rural Resilience. *Journal of Rural Studies*, 52: 1–11.

Biggeri, M. and Ferrannini, A. (2014). *Sustainable Human Development: A New Territorial and People-Centered Perspective*. London: Palgrave Macmillan.

Blay-Palmer, A. (2008). *From Industrial to Sustainable Food Systems*. London: Ashgate.

Bodini, A. (2014). *Food E-commerce: Information-Technology and Marketing Perspective of Its Success factors*. Saarbrucken: Lambert Academic Publishing.

Born, B. and Purcell, M. (2006). Avoiding the Local Trap: Scale and Food Systems in Planning Research. *Journal of Planning, Education and Research*, 26: 195–207.

Brunori, G., Galli, F., Barjolle, D., van Broekhuizen, R., Colombo, L., Giampietro, M., Kirwan, J., Lang, T., Mathijs, E., Maye, D., de Roest, K., Rougoor, C., Schwarz, J., Schmitt, E., Smith, J., Stojanovic. Z., Tisenkopfs, T. and Touzard, J.M. (2016). Are Local Food Chains More Sustainable than Global Food Chains? Considerations for Assessment. *Sustainability*, 8: 1–27.

Carolan, M. (2013). *Reclaiming Food Security*. London: Earthscan.

Clapp, J. (2012). *Food*. Cambridge: Polity Press.

Demetra (2012). Back-to-the-Land Survey Results. Press release, 11 April 2012 [in Greek].

Galli, F. and Brunori, G. (eds.) (2013). Short food supply chains as drivers of sustainable development. Evidence document, developed in the framework of the European Community FP7 project Foodlinks (GA No. 265287), Laboratorio di studi rurali Sismondi.

Gkartzios, M. (2013). Leaving Athens: Narratives of Counter-Urbanization in Times of Crisis. *Journal of Rural Studies*, 32: 158–167.

Haralambakis, E. (2017). What is the Impact of the Crisis in the Economic Status of the Greek Households? A Comparative Analysis of the Two Waves of the HFCS Survey. *Bank of Greece Economic Bulletin*, 45: 37–54.

Heinrichs, C. (2003). The Practice and Politics of Food Systems Localization. *Journal of Rural Studies*, 9: 33–45.

Hellenic Manufacturing (2015). Top 200 agri-food processing firms according to annual sales [in Greek]. Available at: www.inr.gr/?p=a65 (accessed 10 December 2016).

Herslund, L. (2011). The Rural Creative Class: Counterurbanisation and Entrepreneurship in the Danish Countryside. *Sociologia Ruralis*, 52(2): 235–255.

Ilbery, B. and Maye, D. (2005). Alternative (Shorter) Food Supply Chains and Specialist Livestock Products in the Scottish–English borders", *Environment and Planning A*, 37: 823–844.

INEMY (Institute of Commerce and Services) (2016). Annual review of the Greek commerce 2016 [in Greek]. Research study, Hellenic Confederation of Commerce and Entrepreneurship, Athens.

IOBE (Foundation for Economic and Industrial Research) (2016). Food and drink industry: Facts and figures 2015 [in Greek]. IOBE, Athens.

Kneafsey, M., Schmutz, U., Balasz, B. and Trenschard, L. (2013). Short food supply chains and local food systems in the EU. A state of play of their socio-economic characteristics. JRC Scientific and Policy Report 06/2013, European Commission, Luxembourg.

Labrianidis, L., Skordili, S. and Kalogeresis, A. (2004). Entrepreneurship in Rural Greece: Kilkis and Lesvos. In L. Labrianidis (ed.), *The Future of Europe's Rural Periphery*. London: Ashgate, pp. 223–246.

Lang, T. and Heasman, M. (2004). *Food Wars: The Global Battle for Mouths, Minds and Markets*. London: Earthscan.

Morgan, K., Marsden, T. and Murdoch, J. (2006). *Worlds of Food: Place, Power and Provenance in the Food Chain*. Oxford: Oxford University Press.

Mundler, P. and Laughrea, S. (2016). The Contributions of Short Food Supply Chains to Territorial Development: A study of Three Quebec Territories. *Journal of Rural Studies*, 45: 218–229.

Otto, D. and Varner, T. (2005). Consumers, vendors, and the economic importance of Iowa farmers' markets: An economic impact survey analysis. Leopold Center for Sustainable Agriculture, Ames, IA.

Pirog, R., Miller, C., Way, L., Hazekamp, C. and Kim, E. (2014). The local food movement: Setting the stage for good food. Center for Regional Food Systems, Michigan State University.

Renting, H., Marsden, T. and Bank, J. (2003). Understanding Alternative Food Networks: Exploring the Role of Short Food Supply Chains in Rural Development. *Environment and Planning A*, 35: 393–411.

Sage, C. (2002). Social Embeddedness and Relations of Regard: Alternative "Good Food" Networks in South-West Ireland. *Journal of Rural Studies*, 19, 47–60.

Sage, C. (2014). The Transition Movement and Food Sovereignty: From Local Resilience to Global Engagement in Food System Transformation. *Journal of Consumer Culture*, 14(2): 254–275.

Seems, R. (2010). Putting Place on the Menu: The Negotiation of Locality in UK Food Tourism, from Production to Consumption. *Journal of Rural Studies*, 26: 105–115.

Skordili, S. (2013a). The Sojourn of Aldi in Greece. *Journal of Business and Retail Management Research*, 8(1): 68–80.

Skordili, S. (2013b). Economic Crisis as a Catalyst for Food Planning: The Case of Athens. *International Planning Studies*, 13(1): 128–141.

Tregear, A. (2011). Progressing Knowledge in Alternative and Local Food Networks: Critical Reflections and a Research Agenda. *Journal of Rural Studies*, 27(4): 419–430.

Wiskerke, J. (2009). On Places Lost and Places Regained: Reflections on the Alternative Food Geography and Sustainable Regional Development. *International Planning Studies*, 14(4): 369–387.

7 Synergies between localized agri-food systems and short supply chains for geographical indications in Italy

Filippo Arfini and Maria Cecilia Mancini

Introduction

In recent decades, several kinds of local agri-food production models have developed in both EU and non-EU countries in response to globalization (Santini *et al.*, 2013; Lawrence and Almas, 2018). Consumer concerns about food quality play a major role in this trend. The main reasons for buying from local producers include greater variety of products, better safety and traceability, better taste, freshness, less packaging concealing the product, health concerns and environmental awareness of a reduced carbon footprint and fewer chemicals (De Sainte-Marie *et al.*, 2012). In Italy, a model of production which is becoming increasingly popular with consumers and producers is the short food supply chain (SFSC) (INEA, 2014). As SFSCs are developed in a place (*terroir*), their analysis should include geographical space, market relationships and governance. Sometimes, SFSCs are embedded in Localized agri-food systems (LAFSs) and benefit from institutions and coordination structures belonging to the LAFS. LAFS are able to promote new food purchase and consumption patterns, meet quality expectations at fair prices and respond to increasing sensitivities towards local food and rurality (Rytkönen and Hård, 2016). For all these reasons, agri-food system challenges are giving rise to new models of SFSCs which exploit the governance and reputation of LAFS. This is particularly clear for SFSCs of GI products within a LAFS.

The aim of this chapter is to describe how the interactions between different models of local agri-food production systems can produce synergies able to meet new agri-food system challenges and consumer quality expectations. In particular, the SFSC of a product bearing a geographical indication (GI) and its relationships with a LAFS will be analysed. The chapter consists of two parts. The first part presents the theoretical framework underpinning the relationships and synergies between LAFS and SFSCs. The second part presents a case study of Parmigiano-Reggiano (PR) cheese and explores the synergies between a SFSC of this cheese and the LAFS in which such a short supply chain is embedded.

The theoretical framework

The evolution of the concept of LAFS

The first conceptualization of LAFS goes back to the mid-1990s as

> Production and service organizations (agricultural and agri-food production units, marketing, services and gastronomic enterprises, etc.) linked by their characteristics and operational ways to a specific place. The environment, products, people and their institutions, know-how, feeding behaviour and relationship networks combine within a territory to produce a type of agricultural and food organization in a given spatial scale.
>
> (CIRAD-SAR, 1996)

Hence, LAFS can be considered the result of a process of cooperation among companies with common interests, located in a given area, which organize themselves and agree on certain production and marketing norms and rules in order to obtain a competitive advantage over other companies. Competitors can be actual or potential, from within or outside the territory, but they do not adhere to those norms and rules that characterize the LAFS.

Initially, the LAFS production paradigm was approached through the concept of clusters (Porter, 1990), following the idea of spatial proximity of all actors involved in the production model. Muchnik (2009), however, argued that: "the territory of a LAFS is not a continuous space. It is one of belonging, in which a combination of different activities can be carried out in areas that are often physically far apart". It was also agreed that the specificity of the LAFS lies in the features of products, people, institutions and social relations, elements which create/constitute the links between food and territories. Nowadays, researchers consider the relationship between LAFSs and the qualification processes of territorial products as the most important, and collective actions are often developed with a view to obtaining recognition of the origin (Giacomini and Mancini, 2015). In this regard, Muchnik (2009) identifies four aspects of LAFS: qualification of products; co-ordination of stakeholders and collective action; resources management; and dynamics of knowledge. Interaction between these four elements explains the diversity of existing agri-food systems, their evolution, stability and crises.

Lastly, local consumption needs to be recognized as an important component of LAFS for various reasons: (i) the volume of demand helps small farms to make scale economies; (ii) it allows diversification of the food supply; (iii) it gives consumers the chance to base policies on food sovereignty principles; (iv) local economies and traditions are protected; and (v) it strengthens local reputation.

LAFS and local institutions

The leading characteristic of LAFS is a governance which manages relationships between local and non-local actors and institutions, as well as meeting market requirements that have positive effects on the system (Torre, 2000; Rallet and Torre, 2004; Giacomini and Mancini, 2015). In this context, collective action plays a fundamental role in developing strategies to reinforce the competitiveness and sustainability of the whole production system considered as the sum of the territory and its supply chains. Market recognition of local products reflects the collective capacity to define and efficiently manage the combination of natural and human factors. According to this view, collective rules and governance actions should not be considered as a constraint, but rather a condition for guaranteeing sustainability and efficiency for the entire local system (Vandecandelaere *et al.*, 2010).

In this context, the role of local institutions is mainly to strengthen relationships among stakeholders so they can produce public goods and positive externalities which are needed in the construction of quality, and thus increase the level of competitiveness of the local system.

Local institutions can be considered as "all those institutions that represent at local level groups of interest in the economic, social and political pattern" (Vandecandelaere *et al.*, 2010). Torres Salcido and Muchnik (2012) state that local institutions develop actions aimed at reaching agreements and managing key issues related to the local development process in the following areas: technical, institutional, social, market effectiveness, technological improvements, territorial valorization, quality assurance, knowledge transfer and environmental sustainability (see Table 7.1).

LAFS and reputation

Another element which is both a consequence and a distinctive factor of the LAFS production model is reputation. Territorial reputation becomes an economic asset due to the characteristics of the local production system and the role of the local population's consumption model. Belletti *et al.* (2012) argue that local food is "conceived as food with strong roots in a specific geographical place, which gives the product its identity". Consumers within the region are attracted by local food because of its perceived quality. Local foods are delivered to consumers in different ways, such as factory outlets, farmers' markets, community supported agriculture (CSA) and solidarity purchasing groups, as well as restaurants and gastronomic fairs. The efficiency of these systems can be explained by what is known as convention theory (Boltanski and Thévenot, 1991). In these systems, local food becomes a real common good and its value is no longer determined by price or capacity to bring to the table the best food purchased on the market at the lowest price. In what are called "domestic conventions", food ensures support to farmers, ensures positive externalities which

Table 7.1 Areas of intervention of local institutions

Area of intervention	*Objectives*
Institutional	• Establish relationships among territorial and extra-territorial institutions • Comply with national and international regulations
Social	• Promote social inclusion • Promote social cohesion
Market effectiveness	• Establish supply chain relationships • Avoid market failure • Reduce transaction costs • Promote the legal protection of collective names • Generate collective promotion and advertising • Promote consumption of local products
Technological improvement	• Definition of code of specifications • Validation and introduction of new technological patterns • Respect of traditions and local heritage
Territorial valorization	• Management and reproduction of local resources • Promote local marketing
Quality assurance	• Compliance with requisites • Increase levels of trust in product specificity
Knowledge transfer	• Lower technological barriers • Develop professional skills
Environment and sustainability	• Generation of specific environmental actions • Generation of environmental public goods • Management of sustainable policies in rural areas

Source: authors' elaboration of Torres *et al.* (2012).

are both social and environmental and is appreciated by consumers and citizens from the same community. This means that the farmer has an incentive to choose the optimal solution, within a community of consumers and citizens, that can create new attributes for agricultural production and improve the relationship with the environment and social welfare as a whole (Renard, 2003; Renting *et al.*, 2003; Sonnino and Marsden, 2006; Sylvander *et al.*, 2006). Thus, reputation plays a key role in the process of product valorization, helping to guarantee income from local resources, as well as in the economic dimension of the sustainable development process.

SFSCs and LAFS

A specific feature of a SFSC, which generates value along the chain and creates value for producers and the territory, is the level of quality perceived by consumers. The link between local producers and products is indeed a key component of quality expectations and it is the main reason for local consumer willingness to pay, thus creating value for local producers. This implies that territory presents different aspects: (i) it is the place of production; (i) it is the area where the agri-food chain is organised; and (iii) it is the place of consumption.

In this regard, Réviron and Chappuis (2011) consider production and territory at the same time and identify an area consisting of the overlap of two dimensions, territory of production and product value chain, represented by the SFSC. The link between SFSC and territory is usually strong but not explicit, although when the output of a SFSC is a GI product the link is both strong and explicit.

When embedded in a LAFS, SFSCs can benefit from governance organizations which oversee quality management and the relationship with public and private institutions in a collective manner. This is evident in those SFSCs that deal with GI products (GI-SFSCs) where the governance action in the territory and in the chain is developed by inter-branch organizations (Giacomini *et al.*, 2011; Arfini and Mancini, 2013; Giacomini and Mancini, 2015).

It can be argued that local institutions are hybrid organizations, as they represent the collective interests of individual producers involved in an agri-food chain within a territory. In fact, according to the literature on collective action (Vandecandelaere *et al.*, 2010; Réviron and Chappuis, 2011), the production of local products involves the participation of different types of stakeholders operating within and outside the production area (see Table 7.2), where only some of them are directly involved in the value creation process while others operate at the local level. So, collective action generated by inter-branch organizations and local institutions has implications at the value chain and territorial levels.

Table 7.2 Stakeholders involved in the value creation process and territorial relationships

		Food chain	
		Inside the agri-food chain	*Outside the agri-food chain*
Territory	**Inside the territory**	• Producers • Processors • Local consumers • Local citizens • Policy makers	• Local public authorities • Local intermediate institutions • Research and extension • Other economic activities
	Outside the territory	• Other producers • Other processors • Distributors • Retailers • Consumers	• Other public authorities • Other institutions • Research centres • Consumer associations

Source: authors' elaboration based on Vandecandelaere *et al.* (2010).

A classification of LAFS

The different features of the agri-food supply chains in the territory define three types of LAFS (see Figure 7.1):

1 *Closed systems*: local agricultural outputs are processed by local producers, local food industries (mainly SMEs) and purchased mainly by local consumers.

This LAFS model is characterized by the strong tie between the agricultural phase, processing companies and local consumers. This tie impacts on product quality, firm structures, market strategies and social relationships. As a consequence, management of the local environment is the key aspect of a LAFS; it contributes to managing the quality of the inputs, acts towards guaranteeing a certain volume of production and the reproduction of natural resources and reinforces the image (and reputation) of the system as a whole. In this production model, local resources are important. They consist of environmental factors such as characteristics of land and water, as well as features of the history and environmental specificity of the region, e.g. biodiversity, animal breeds and local tradition.

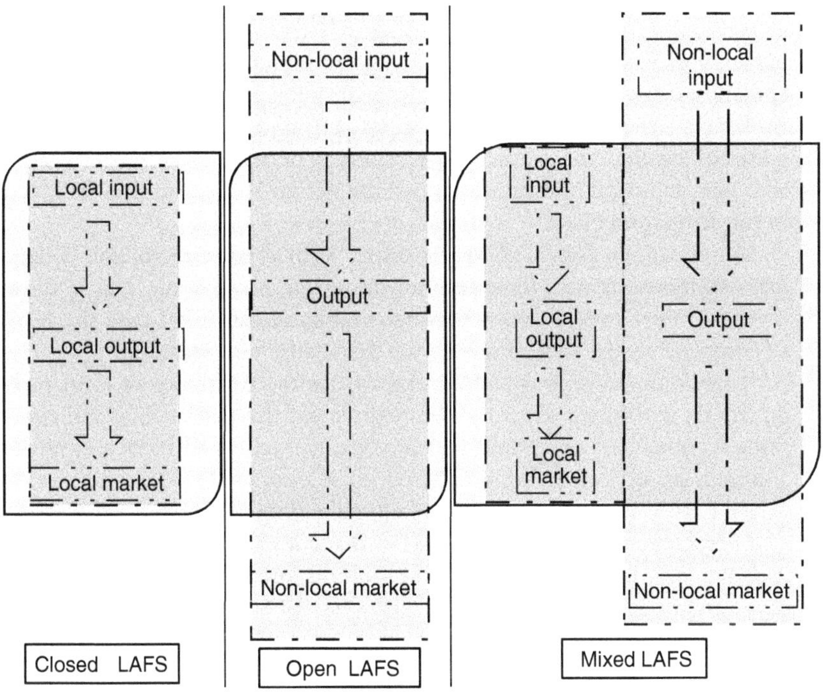

Figure 7.1 Relationships between agri-food chains and LAFS.

Source: authors.

Reputation in a closed system is thus linked to local characteristics and the know-how developed over time by local consumers in making traditional foods and recipes which have become famous outside the area. It is the combination of these two aspects (production system and consumption style) which makes reputation an asset for products and producers in the area.

2 *Open systems*: the supply basin of farm input products and/or the product final market is outside the LAFS.

In some LAFS, the downstream parts of the supply chains are outside the LAFS and the output is sold onto larger markets (Becattini, 1989). Often, the higher the reputation, the further away are the consumer markets. The result is the presence of agri-food chains that can include different types of actors, where only some of them are operating within the territory. In fact, processing companies from the LAFS have their historical roots in the area and have developed efficient and effective marketing strategies towards global markets. These companies have often generated global food chains with effective management of both production systems and consumer relationships. Often, these companies become multinational enterprises with factories all over the world and headquarters in the place of origin.

Inputs may also come from outside the LAFS. In this case, the link between the local agriculture system and processing industries is weak, as in the case of the largest agri-food and GI food chains. For such companies, agricultural inputs are not a constraint and their strategy is to buy agricultural commodities at the right quality for processing at lower cost. In this case, the links between the supply chain and the territory are concretized through the labour force, cultural heritage and skills, research activities, logistical infrastructures and the network of enterprises involved in the same food chain.

Moreover, in "open" LAFS models, local companies might benefit from connections with local and non-local research systems, which allow them to innovate and follow new technological paths that raise the level of competitiveness without losing the links with local traditions.

Reputation in open systems comes from the capability of local producers to innovate and to be competitive on the market but also from skills accumulated over time and governance action to reduce conflicts and transaction costs in business activities at local level.

3 *Mixed LAFS systems*: coexistence of closed and open LAFS.

These systems are characterized by the coexistence of both "closed" and "open" LAFS models where outputs are sold on either local, domestic or global markets. Nevertheless, SFSCs play a key role in meeting local demand and building territorial reputation, especially where markets include tourists attracted to the area by food specialities and the environmental features of the LAFS. Moreover, coexistence strengthens all the variables that characterize the development process of the LAFS, including reputation.

Reputation becomes an asset for all the firms involved in the food production system if it is associated with local products rooted in the area and with the name of the region, which often bear GI recognition. In this case, the name of the region becomes a brand, not only for industries involved in the GI products but also for the entire food sector and local companies, and it carries a message of quality. An important phenomenon is the "spill-over reputation" effect which gives a central place to the reputation of the actors in the management of the development process or crisis situations (Mayer, 2006; Yu and Lester, 2008; Giacomini *et al.*, 2010).

In addition, in order to preserve and exploit the reputation of the region, local institutions cooperate and take advantage of "cross fertilization" generated by the environment of the LAFS. SFSCs producing GI products particularly benefit from consumer perception of quality as described in domestic convention theory as well as the LAFS environment (see Figure 7.1).

The case study: the short supply chain of Parmigiano-Reggiano in Parma Province

The Parma LAFS

Italy is well-known all over the world for the quality of its agri-food products. The agri-food industry covers all regions of the country, but some are more highly specialized. An example is the Emilia-Romagna region and the Province of Parma. The Parma Province can be considered an example of a mixed LAFS for the following reasons:

1 It has a very intensive agricultural sector.
2 It has large agri-food companies in the sectors of tomato (50 per cent of Italian tomatoes are processed in the area of Parma), sugar (one of the few factories in Italy that still processes sugar-beet is in Parma Province), pasta and baking (Barilla is the leading company in the sector), dairy products (Parmalat is still one of the most important dairy sector companies in Italy as well as worldwide) and pork meat (with large slaughterhouses).
3 It has SMEs that produce GI products: three Protected Designation of Origin (PDO) products (Parmigiano-Reggiano, Prosciutto di Parma and Culatello di Zibello) and two Protected Geographical Indication (PGI) products (Salame di Felino and Fungo di Borgotaro).
4 It has innovative SFSCs, including farmers' markets and CSA schemes. Some of the products listed are sold on global markets but others are sold mainly in the area of production to domestic consumers or tourists attracted by the reputation of Parma's LAFS.

The agri-food chains of Parma's LAFS benefit from governance institutions such as GI consortia, producer associations and inter-branch organizations.

These institutions act as *third party* institutions and support the agri-food supply chains directly, by developing a specific governance strategy for each member, as well as indirectly, for the area of production (Giacomini *et al.*, 2011). In addition, other public institutions, including research centres and chambers of commerce, interact with private institutions and agri-food companies with the aim of supporting them and increasing their competitiveness on local, national and international markets. This is helped by spatial proximity and personal relationships between members (Mancini and Arfini, in press). Other factors associated with spatial proximity include the generation of positive externalities such as reputation spillover between different food products from the Parma area (Giacomini *et al.*, 2010).

In brief, Parma's LAFS includes: (i) links between material and immaterial goods; (ii) the presence of common objectives and strategies related to quality and environmental policies, a reduction in transaction costs and raising the competitiveness of all stakeholders in the LAFS; and (iii) the coexistence of local and global agri-food chains. The whole system benefits from the fact that all these companies are located in the same territory. This facilitates industrial relations, including workforce management, and generates externalities, such as reputation, which makes the LAFS competitive on both domestic and international markets.

The PR production system

Parma is one of the provinces specified in the Parmigiano-Reggiano (PR) PDO code of specifications (see Figure 7.2) and the PR production system is part of the Parma LAFS. PR cheese significantly contributes to the reputation of the Parma area; its production system is responsible for several positive externalities due to the presence of milk producers in mountain areas and the use of low environmental impact production techniques. For this reason, the PR production system is considered an important element in the sustainability of the whole Parma LAFS (De Roest, 2000).

The institution governing the PR production system is the PR Consortium (CFPR) which supports firms in production and sales as well as relationships with Italian and international institutions.

The actors on the PR supply chain comprise:

1 *Farmers*. There are three different types: (i) farmers who deliver milk to cooperative dairies they are members of; (ii) farmers who sell their milk to non-cooperative (small-scale or industrial) dairies; and (iii) farmers who process the milk they produce in their own dairies.
2 *Dairies*. There are also three different types of dairy: (i) farm-owned; (ii) cooperative; and (iii) non-cooperative (small-scale or industrial). A common feature of all PR dairies is that they are single-product businesses, as the milk is used exclusively for production of the cheese.

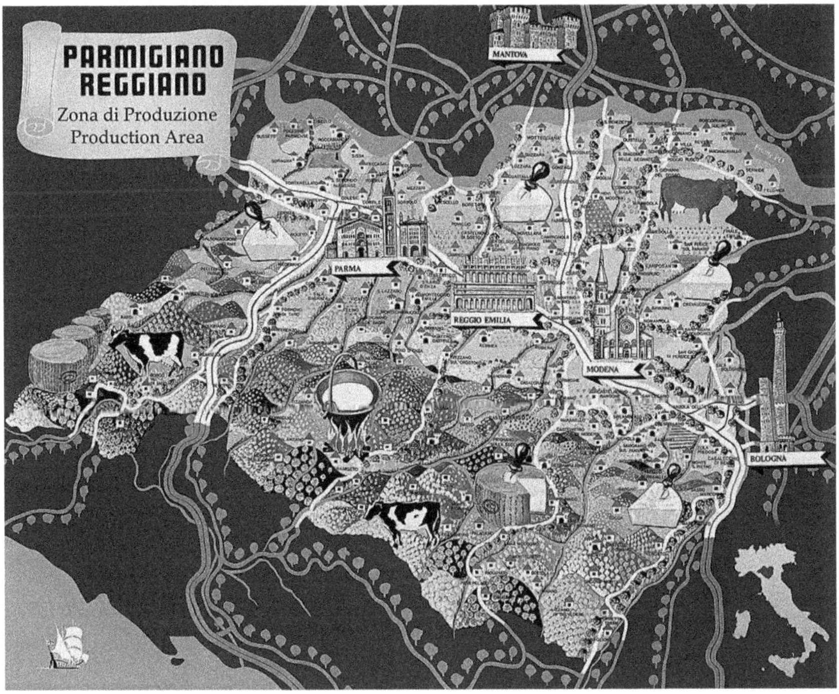

Figure 7.2 PR area of production as defined in the code of specifications.
Source: authors.

Cooperative dairies are the main, if not the exclusive, source of income for member farms.

3 *Wholesaler-agers.* They play a key role in the PR supply chain, as they mature the cheese for the second and third year and develop sales strategies to place it on the market. They set cheese price, develop competitive strategies and more than any other actor take advantage of the promotional brand campaigns run by CFPR.
4 *Research centres (including the five universities of Emilia-Romagna) and technical laboratories.* These provide technical assistance for farmers and dairies and promote innovation processes.
5 *Private and public institutions.* Institutions (e.g. CFPR, chambers of commerce) take part in the governance of the PR supply chain as well as the whole area in regard to economic and political matters.

Some of these actors, such as research centres, are located in the area of Parma but also act outside the PR supply chain, as they support the entire milk sector nationwide. Others act in the PR supply chain but are located outside the province (see Figure 7.3).

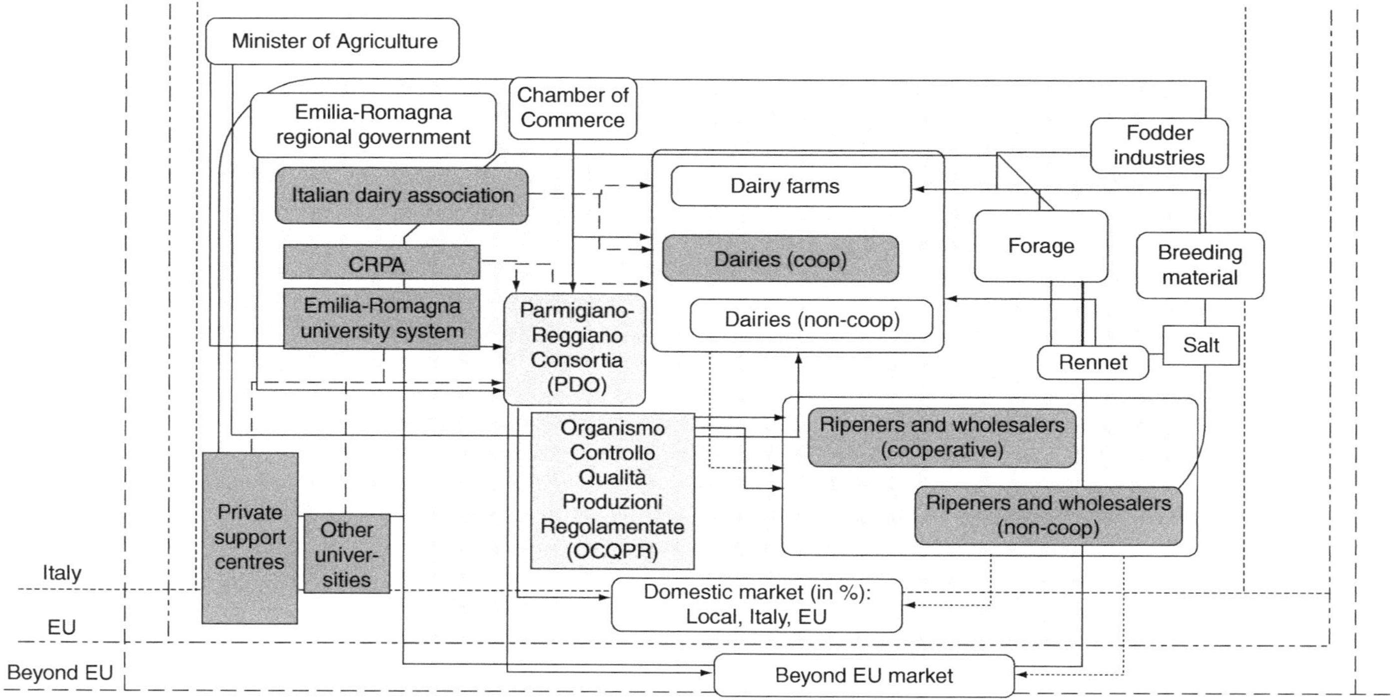

Figure 7.3 The PR production system.

Source: authors.

The PR supply chain as a whole is governed by the CFPR, which plays the role of a third-party institution and has relationships with the actors in the Parma LAFS. The CFPR is the only institution governing the entire supply chain and its decisions are mandatory for all members. The CFPR has three main functions: (i) to guarantee cheese quality over time; (ii) to protect PR producers against unfair competition; and (iii) to promote this PDO cheese on the market. Producers are free to determine their volume of production within the constraints of the milk quota set by the CFPR and their sales strategy.

The length of the financial cycle reflects the long maturation period of the product and has led to a clear division of tasks between milk producers, dairies and wholesaler-agers. At the same time, close integration has developed along the supply chain, both formally among farmers and the dairies, and informally, between dairies and wholesaler-agers. This was shown in Arfini *et al.* (2006), who ascertained that, after one year of maturation, almost 50 per cent of cheese is sold to the same wholesaler-ager. It should be noted that PR is mainly sold out of the territory of Parma on national and international markets, while only a small percentage of cheese (about 1 per cent) is sold in the province itself through different distribution channels, including the short supply chain.

One of the main characteristics of this supply chain is that most dairies, especially cooperative dairies, age the cheese in maturing warehouses in as short a time as possible, selling their output to wholesaler-agers as soon as the product and market conditions allow. So the cheese is sold in batches at 12 months, as soon as it has been PDO and CFPR branded, to wholesalers with whom they have consolidated relationships of trust as stable sales partners (Arfini *et al.*, 2006). This strategy is aimed at reducing marketing costs and focussing dairy businesses on milk transformation. Wholesalers age the cheese for the second year, and then sell it on to other wholesalers or directly to retail channels using their own marketing strategy.

The main implications of this business model are: (i) dairies lose individual identity, as wholesalers sell the cheese under their own brand and consumers are unable to recognize the quality characteristics of different producers; and (ii) dairies lose market power as wholesalers can impose prices, especially where there is excess production. These characteristics make PR market trends more typical of commodity goods than speciality goods, with structural cyclical crises where high production generates low market prices and low production generates high market prices (Arfini *et al.*, 2006).

The PR short supply chain in Parma LAFS

It is widely recognized that PR market crises are due to structural weaknesses. Past attempts to modernize the market structure have been unsuccessful because of the length of the supply chain and the number of actors involved, and also because producers need to sell the cheese after 12 months of ageing. It has never been possible to introduce market strategies that create market distinctiveness for dairies and increase their market power. However, the most recent challenges have led the CFPR to develop a new marketing strategy aimed at boosting PR sales directly at the dairy outlets.

This strategy is underpinned by two current trends: (i) the fall in the number of dairies (from 733 in 1993 to 335 in 2016) and the rise of production capacity (De Roest and Menghi, 2013) which allows dairy managers to directly manage sales and marketing strategies, including direct selling part of their cheese production; and (ii) the increasing popularity of buying food from short supply chains as more informed consumers look for better quality at a reasonable price.

The volume of PR sold through dairy outlets is a small percentage of the whole production, but the strategy is important because it improves the short-term financing capacity of dairies and provides a closer link between producers and consumers.

The CFPR's governance on this strategy has three strands: (i) the promotion of dairies with outlets on the official PR website; (ii) standardization of the outlet format; and (iii) outlet staff training.

With regard to the first strand, the CFPR website (www.parmigiano-reggiano.it) is rich in information for members and consumers; it provides information on the production system, the code of specifications, the certification system and recipes that use PR as an ingredient. It is one of the main communication tools for PR producers and promotes dairies which sell cheese directly. It provides, in a user-friendly manner, an image of quality and trust for the whole system while at the same time providing practical information. In particular, the website: (i) shows the locations of the dairies; (ii) specifies whether the dairies sell on-site or by e-commerce; and (iii) provides consumers with contact details for outlets.

In the Parma LAFS, CFPR data for 2012 show that of the 180 dairies in business, 49 had a website, 66 had a dairy outlet and eight sold to consumers directly through e-commerce (www.parmigianoreggiano.it). Of the 66 dairy outlets, seven were also equipped to sell through e-commerce and only 36 only had their own website. So, there are different strategies in developing a short supply chain, which vary according to the management characteristics of each dairy and the technological and interpersonal skills of staff.

The second strand of the CFPR's strategy is the design of a dairy farm outlet standard. Parma LAFS institutions played a role here and the store layout was designed in collaboration with local research centres. The aims were to generate a single image of the outlets from the consumer's perspective and to reduce the costs of promotion and marketing innovations in each outlet. The process was driven by the CFPR and has led to standardization of: (i) outlet format; (ii) training and uniforms for staff, and free gifts; and (iii) methods and procedures for orders, shipments and delivery.

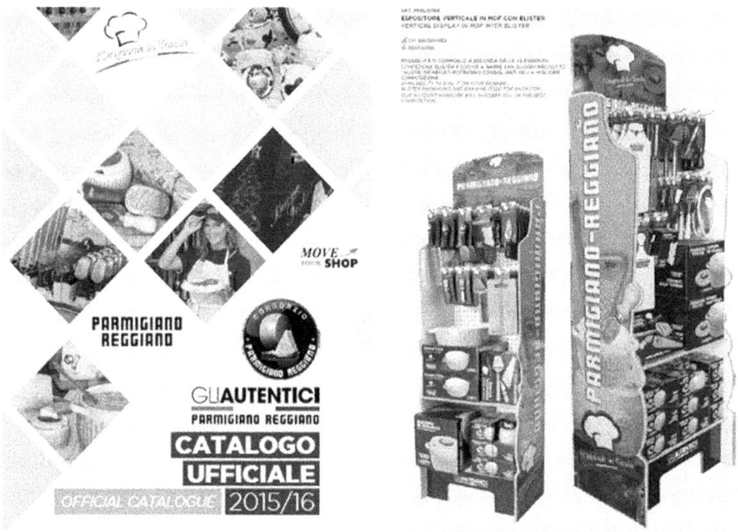

Figure 7.4 PR catalogue and merchandising.

Sales strategy in outlets is thus standardized to a great extent, but producers are free to sell any food products they wish. The following goods are usually on sale at outlets, as well as PR:

- other cheeses (ricotta, other GI cheeses);
- local GI products (PGI Salame di Felino, other types local salami, PGI porcini mushrooms from Borgotaro);
- non-local GI products (PDO olive oil);
- CDO (Controlled Designation of Origin) wine from the Parma area;
- traditional pasta (local and non-local);
- Parmigiano-Reggiano knives, graters, etc. and recipe leaflets;
- other products (food and non-food) according to the location of the dairy.

Staff wear a standardized, specially designed uniform. The CFPR has produced a catalogue of gifts which can be displayed in the shops (Figure 7.4). PR is often bought as a gift, so accoutrements such as special cheese knives are often sold with it; the gift-wrapping also comes from the CFPR. Sales and marketing activities are supported by technologies developed by the CFPR for logistics such as hardware, software and packaging tools.

The third strand of CFPR strategy is training to provide dairy staff with shop management skills in various areas: safety requirements, the traceability system, fiscal aspects, marketing strategies, labelling, shipments, other regulations, etc. Staff are also trained to show tourists around the cheese factory and describe the production process and its artisanal and natural features. Training is a very important component of the short supply chain strategy as dairy personnel usually have little formal education and experience in direct sales.

The short supply chain strategy for PR dairies has produced direct and indirect effects for the PR system. Direct effects at dairy level are: an increase in cash flow, a rise in added value, greater contractual power over retailers and increasing customer loyalty through direct sales. Indirect effects are on the socio-economic sustainability of the Parma LAFS. The strategy has given producers, especially those in remote areas, a new economic basis by boosting endogenous rural development and promoting new forms of territorial marketing connected to enogastronomic tourism. Empirical observation of the evolution of the PR dairies in the Parma LAFS shows that over six years of economic crisis (2007–2012), all dairy factories with direct sales were still active, while 30 per cent of those without outlets closed down (Mancini and Arfini, in press).

Conclusions

Emerging challenges in agri-food systems in recent years have impacted on behaviours and strategies in both consumption and production. One of the effects is that SFSCs have appeared or, more precisely, reappeared, so that today consumers and producers can once again meet one another, exchange information and, to certain extent, jointly decide on strategies for purchasing and production.

SFSCs can be a model which meets the requirements of consumers, optimizing quality and price, and giving producers a large share of added value. But it is important to note that the SFSC model is based on the specific nature of the product, the social environment and the organization of the supply chain.

The most frequently described SFSCs in the literature are farmers' markets and solidarity purchasing groups. In these models, the products are mainly fresh produce bought directly from farmers that have no downstream links to other companies. But short supply chains of processed foods, such as cheese, are different. These products are often characterized by a high level of reputation over the whole production area, which is an attraction factor for both local and non-local consumers. For processed food, local institutions put producers and consumers in contact and their activities directly benefit supply chain producers and indirectly benefit other firms in the LAFS.

As we have seen, LAFS themselves also differ. There is a continuum between completely closed systems and open systems. Closed systems are characterized by local production and consumption and there is little interaction with other territories, while processing only is carried out in the area when the LAFS is open. Through the governance of LAFS institutions, both closed and open systems are able to meet new consumer and producer requirements. Mixed LAFS are a particularly interesting case; they include industrial production systems, typical of open LAFS, as well as artisan and semi-industrial systems, typical of closed LAFS, and product reputation is closely linked to the reputation of the area. Context underpins new types of SFSCs: the Parma LAFS is emblematic. As a mixed LAFS, it has succeeded, through its local institutions, in providing dairies with a short supply chain for Parmigiano-Reggiano that is able to strengthen producers' resilience to agri-good system challenges.

References

Arfini, F. and Mancini, M.C. (2013). Local institutions and territorial competitiveness in the case of Parmigiano Reggiano localised production system. Presented at the 2nd AIEAA Conference *Between Crisis and Development: Which Role for the Bio-Economy*, 6–7 June 2013, Parma, Italy.

Arfini, F., Boccaletti, S., Giacomini, C., Moro, D. and Sckokai, P. (2006). Case study: Parmigiano Reggiano. European Commission, Brussels.

Becattini, G. (1989). Riflessioni sul distretto culturale marshalliano come concetto socio-economico. *Stato e mercato* 25, pp. 111–128.

Belletti, G., Casabianca, F. and Marescotti, A. (2012). Local Food Quality and Local Resources. In: F. Arfini, M.C. Mancini and M. Donati (eds.), *Local Agri-Food Systems in a Global World: Market, Social and Environmental challenges*. Newcastle upon Tyne, UK: Cambridge Scholars Publishing,

Boltanski, L. and Thévenot, L. (1991). *De la Justification. Les Économies de la Grandeur.* Paris: Gallimard.

CIRAD-SAR (Centre de Coopération Internationale en Recherche Agronomique pour le Développement-Systèmes Agro-alimentaires et Ruraux) (1996). *Systèmes Agroalimentaires Localises: Organisations, Innovations et Développement Local.* Montpellier: CIRAD-SAR.

De Roest, K. (2000). *The production of Parmigiano-Reggiano cheese: The force of an artisanal system in an industrialised world.* PhD thesis, Wageningen University.

De Roest, K. and Menghi, A. (2013). Costi di produzione e di trasformazione del latte in Emilia-Romagna. CRPA (Research Centre on Animal Production), Reggio Emilia, Italy.

De Sainte Marie, C., Amiaud, B., Magda, D., Plantureux, S., Agreil, C., Mestelan, P., Boussou, V. and Mougey, T. (2012). Promouvoir la valeur agri-écologique des prairies: le concours agricole national des prairies fleuries. Presented at *Prairies Permanentes: De Nouveaux Atouts pour Demain*, Paris, 3–4 April 2012.

Giacomini, C. and Mancini, M.C. (2015). Organisation as a Key Factor in Localised Agri-Food Systems (LAFS). *Bio-based and Applied Economics* 4(1), pp. 17–32.

Giacomini, C., Arfini, F. and Menozzi, D. (2010). Processi di Qualificazione e Spillovers: il Caso del Prosciutto di Parma Dop. *La Questione Agraria* 3, pp. 55–81.

Giacomini, C., Arfini, F. and De Roest, K. (2011). Interprofession and Typical Products: The Case of Parmigiano Reggiano cheese. *Sviluppo Locale* 15(37/38), pp. 125–150.

INEA (2014). *L'agricoltura in Italia. Multifunzionalità, Diversificazione e Servizi.* Roma: MIPAF.

Lawrence, G. and Almas, R. (2018). *Globalisation, Localisation and Sustainable Livelihoods.* London: Routledge.

Mancini, M.C. and Arfini, F. (in press). Producers Strategies for an Italian Protected Designation of Origin Product between Economic Crisis and Quality Consumers' Expectations – the Short Supply Chain of Parmigiano Reggiano. *Journal of Depopulation and Rural Development Studies*.

Mayer, K.J. (2006). Spillovers and Governance: An Analysis of Knowledge and Reputational Spillovers in Information Technology. *Academy of Management Journal* 49(1), pp. 69–84.

Muchnik, J. (2009). Localised agrifood systems: Concept development and diversity of situations. Presented at the Annual Meetings of the Agriculture, Food, and Human Values Society and the Association for the Study of Food and Society, State College, Pennsylvania, 28–31 May 2009.

Porter, M.E. (1990). *The Competitive Advantage of Nations*. New York: The Free Press.

Rallet, A. and Torre, A. (2004). Proximité et Localization. *Economie Rurale* 280(1), pp. 25–41.

Renard, M.C. (2003). Fair Trade: Quality, Market and Conventions. *Journal of Rural Studies* 19(1), pp. 87–96.

Renting, H., Marsden, T.K. and Banks, J. (2003). Understanding Alternative Food Networks: Exploring the Role of Short Food Supply Chains. *Rural Development, Environment and Planning* 35, pp. 393–441.

Réviron, S. and Chappuis, J.M. (2011). Geographical Indications: Collective Organization and Management. In: E. Barham and B. Sylvander (eds.), *Labels of Origin for Food. Local Development, Global Recognition*. Cambridge, MA: CAB International, pp. 45–62.

Rytkönen, P. and Hård, U. (eds.) (2016). Challenges for the new rurality in a changing world. Proceedings from the 7th International Conference on Localized Agri-Food Systems, Södertörn University, Stockholm, 8–10 May 2016.

Santini, F., Guri, F., Aubard, A., Psaltopoulos, D., Read, R. and Gomez y Paloma, S. (2013). EU island farming and the labelling of its products. JRC Scientific and Policy Report, European Commission, Luxembourg.

Sonnino, R. and Marsden, T. (2006). Beyond the Divide: Rethinking Relationships between Alternative and Conventional Food Networks in Europe. *Journal of Economic Geography* 6, pp. 181–199.

Sylvander, B., Belletti, G., Marescotti, A. and Thévenod-Mottet, E. (2006). Establishing a Quality Convention, Certifying and Promoting the Quality of Animal Products: The Case of Beef. *European Association for Animal Production*, 118(61), pp. 61–82.

Torre, A. (2000). Faut-il avoir Confiance dans les Techniques de Production de la Confiance, Liens de Proximité et Stratégies des Acteurs. In: R. Laufer and M. Orillard (eds.), *La Confiance en Question*. Paris: L'Harmattan.

Torres Salcido, G. and Muchnik, J. (2012). Globalization/Fragmentation Process: Governance and Public Policies for Localized Agri-Food Systems. In: F. Arfini, M.C. Mancini and M. Donati (eds.), *Local Agri-Food Systems in a Global World. Market, Social and Environmental Challenges*. Newcastle upon Tyne, UK: Cambridge Scholars Publishing, pp. 97–116.

Vandecandelaere, E., Arfini, F., Belletti, G. and Marescotti, A. (2010). *Linking People, Places and Products*. Rome: Food and Agriculture Organization.

Yu, T. and Lester, R.H. (2008). Moving beyond Firm Boundaries: A Social Network Perspective on Reputation Spillover. *Corporate Reputation Review* 11(1), pp. 94–108.

8 Re-embedding Greek Feta in localities

Cooperation of small dairies as a territorial development strategy

Theodosia Anthopoulou and Dimitris Goussios

Introduction

Geographical indications (GIs) are place-based names that convey the cultural and historical identity of local agri-food products. GI schemes, especially Protected Designation of Origin (PDO), were conceived to protect the specific quality and the intellectual property of rural communities that have over time created these identity products. They are therefore considered as an efficient policy tool to leverage wider economic, social and environmental benefits to rural areas, especially the mountainous and less-favoured ones.

In the case of Greek Feta cheese, the final decision made by the European authorities that granted Greece PDO rights in 2005, after ten-year litigation with other European countries, gave new impetus to both the domestic and the international market. Nevertheless, the empowering of the position of the Greek dairy industry in the global market created conflicting fragmentation processes within the production chain. What actually happened was that the implementation of the PDO scheme strengthened large firms while generating conditions of market exclusion for small dairies operating at a regional/local scale. As also noted by several scholars, the PDO label seems to be insufficient in ensuring fairness and sustainability for local traditional food systems within the globalized market (Bowen, 2010; Dupuis and Goodman, 2005, Torres *et al.*, 2012; Marescotti and Belletti, 2016).

In these global–local dynamics, the main question that arises is: How can a globalized food such as Greek Feta cheese be re-localized so as to reproduce the specific territorial resources of the agropastoralist areas and define a territorial development strategy? In particular, two key issues emerge: (i) To what extent are certification schemes efficient in reproducing locally embedded cultural and natural resources (i.e. the *terroir*) and contributing to socially and economically fair GI production systems at the local level?; and (ii) How can we shift from the governance of the local dairy system (sectoral/economic approach) to a territorial governance strategy (i.e. territorial approach to local/rural development) (Campagne and Pecqueur, 2014: 173–204; Muchnik *et al.*, 2008: 516)?

In this chapter[1] we argue that the specification process (creation of a regional Feta label) within the nationally defined PDO area of Feta cheese (i.e. the product is already protected for its features against counterfeiting) strengthens small dairy territories by activating territorial assets (pastures and natural flora, locally adapted breeds, cheesemaking techniques, informal kin and ethnocultural networks, etc.). This further implies the empowerment of small dairy enterprises and an improvement in the position of actors involved in the supply chain (stockbreeders, fodder producers, local food processors, etc.). Working in the field in Thessaly region (Central Greece) since 2012, our main objective was to create an origin identity Greek Feta of distinctive quality (within the national PDO) vis-à-vis a PDO Greek Feta of industrial standard quality intended for the global market – an endeavour which was encompassed within a territorial development strategy. This chapter outlines the procedures for: (i) creating a regional label (under the "Terra Thessalia" brand name) with the participation of local actors involved in the value chain (bottom-up approach); and (ii) developing a territorial governance strategy that aimed to make this endeavour viable.

Localized agri-food systems, territorial specificities and cooperation in the era of globalization

In recent years many studies have pointed to the emergence of an alternative food system, including both consumption and production aspects, in response to globalization and the industrialized agricultural model. The quality turn of consumption creates new opportunities for the local agri-food sector, through quality differentiation based on social, environmental and territorial specificities, and consequently new opportunities to strengthen the position of farmers, processors and SMEs in the value chain (Rytkönen *et al.*, 2018).

Drawing on the French conceptualization of localized agri-food systems (SYAL – *Systèmes Agroalimentaires Localisés*), a growing, relevant literature is shedding light on the complex relationships between the agri-food system and the territory.[2] Particular attention is given to consumer behaviour and its role in the valorization of local food products through purchasing choices increasingly driven by distinctive quality. Localized agri-food systems are perceived as being embedded in a concrete territory which brings specific attributes to local food. These are associated with three key factors involved in a territorial context: (i) strong ties with local natural and biophysical resources used in the production process; (ii) cultural ties, such as history, local knowledge, traditional practices of production and consumption processes; and (iii) the collective dimension of sharing common knowledge, traditions, the origin's reputation, patrimony and, furthermore, what is considered as "territorial intelligence" (Muchnik, 2006; Belletti *et al.*, 2012). That said, the territory is not just a physical support for economic activities but also a space of identity reference, a kind of natural and cultural matrix that gives shape to local activities and enables their articulation in a "world economy". Under the lens of the SYAL approach, a territory "forms a system" including the organization, functioning and interactions of various activities whether they are productive, social or cultural. In this way, the territory becomes an activator of local (latent) resources (Muchnik *et al.*, 2008) and thus a lever for local/regional development.

In a globalized world, where asymmetrical power relations alongside a value chain comprised of very diverse local and extra-local actors (farmers, processors, retailers, governmental officers, etc.) create the risk of appropriation of territorial resources by the major players of the supply chain, the question of coordination of the involved territorial actors and of governance is critical. Beyond geographical proximity, social and cultural proximity and the sense of belonging and sharing a common cultural heritage may also forge cooperation aimed at promoting territorial goods of specific quality (Enriquez-Sanchez *et al.*, 2017). Muchnik *et al.* (2008) argue, in the light of the SYAL approach, that coordination between local actors and their organizations based on common identity references enables the development of "territorial solidarity" as well as the emergence of common rules that could facilitate projects and collective actions. In this way, the "territorial solidarity" and the organization of a localized agri-food system (SYAL) allow the strategies and economic activities of individuals to be articulated under a collective project meaningful for both locals and non-locals, but with common identity references in relation to a territory.

The SYAL approach puts forward two major articulated steps. The first refers to the identification and qualification of the product as a GI, while the second focuses on the activation of territorial resources, both encompassed in a "virtuous circle of origin-linked quality" to boost sustainability of the system (Vandecandelaere *et al.*, 2009). This virtuous circle corresponds to a value creation process along four main stages: (i) product identification by local stakeholders motivated to reveal local specificities and initiate a collective process (identification); (ii) qualification of the GI specificities (defining a code of practice); (iii) remuneration through the market as recognition by consumers of the prime quality of the product (price differential); and (iv) reproduction of territorial resources and, consequently, of the GI product (environmental aspects and fair redistribution of added value to the actors involved). The whole process relies on actors' organization and cooperation through collective action (ensuring territorial governance) in order to manage the link between product quality and territory and maintain sustainable GI projects.

Drawing on the SYAL approach and the "virtuous circle" of a GI product process, this chapter aims to explore how the territorial specificities to a PDO product of international appeal can be re-localized and re-ascribed.

Re-embedding Greek Feta in traditional small dairy territories under the PDO regime

The complexity of Feta cheese is linked to both its local and its global dimension: it is a culturally emblematic Greek cheese which is also renowned in the global market. The high visibility of Feta in the international market led to the famous "Feta wars" which broke out after the first approval (1996) of its designation as a Greek brand under a European Directive (EEC No. 2081/92). The key argument of the countries that denounced the PDO status of Greek Feta (Denmark, Germany and France) was that it is a "generic term" that is not unique to Greek origin, as they had also been producing it for many years (an imitation "white cheese", made mainly from cow's milk, not sheep's milk). In 2005, after an extensive investigation conducted by the European Commission on the origins and history of the cheese, Feta was definitively registered as a Greek PDO product. It is considered to represent an integral part of the everyday diet and culinary culture of the Greek people, and produced in delineated areas of the country from sheep and goat's milk, according to specific codes of practice, as a nationwide PDO.

The opening of new markets, both domestic and foreign, and the significant increase in demand for the Greek product was however mainly exploited by large industrial dairies, by virtue of their efficient technical and organizational capacity to adapt. They had the requisite infrastructure, which enabled them to collect milk across a wide territorial radius, as well as modern technologies, which facilitated economies of scale through standardization processes of production that could respond to national and international market requirements (industrial quality standards, sizeable quantities). These dynamics led to spatial concentration of industrial dairy units in high-density livestock areas (e.g. Thessaly, Epirus, Central Macedonia) with a tendency to monopolize supplies of raw material, i.e. breeders of small ruminants.

On the other hand, small and medium artisanal dairies that have remained attached to local socio-productive systems operate mainly through personal networks both at an upstream (breeders) and a downstream (traders) level. They are anchored in local social, cultural and ecological resources (extensive pastures, local breeds, ethnocultural pastoral groups, traditional cheese making practices) within their small territory, which is traditionally specialized, historically constructed and geographically defined by the milk collection radius. Loyalty and a sense of belonging to the pastoralist community through personal networks contribute to the smooth functioning of small traditional dairies whose competitive advantage is a specific place-based quality in the absence of economies of scale. Nevertheless, the lack of cooperation among artisanal dairies within their small territory has configured socio-spatial enclaves of traditional breeding systems inside wider dairy territories without synergies. Yet this kind of "heritage repository" keeps local specificities active within PDO standardized quality as a cultural backdrop fuelling the public imaginary of Greek Feta. At present, an important part of the added value is captured at the end of the value chain – that is, the distribution and retail networks – who exert downward pressure on prices back upstream (first cheese processors and then breeders) in order to reduce production costs (Anthopoulou and Goussios, 2015).

It should also be noted that in this polarized and conflicting market situation, breeders do not benefit from a differential price according to the quality of the milk intended for Feta PDO production. In the absence of collective arrangements for the negotiation of farm prices (the lack of cooperatives, farmer groups and dairy inter-professional structures), the producer's price and time of milk collection are fixed by individual, mainly verbal, agreements between sheep and goat breeders and processors. In these contrasting dynamics, we see a decreasing trend in the number and size of sheep and goat livestock herds, especially in extensive mountainous grazing systems, including transhumance. Likewise, there is a tendency of small and very small dairies to disappear, thus risking the loss of cultural heritage resources in territories with a strong pastoralist identity. Yet it is the small producers (breeders, dairy farmers, cheese processors, local creameries, etc.) and rural communities that ensure the reproduction of agro-pastoralist systems and traditional techniques, contributing to the socio-cultural and environmental sustainability of mainly mountainous and semi-mountainous Greek areas.

Drawing on a diagnostic survey and the elaboration of a territorial strategy for developing quality dairy products in the Thessaly region (Central Greece), the remainder of this chapter focuses on three main issues:

1 whether the activation of territorial assets gives perspective in creating differentiated cheese of origin-based quality at regional level under national GI regulations – i.e. Feta PDO coming from traditional small dairy territories of the Thessaly region, which in turn consolidates this same PDO national label;
2 how a differentiated Greek Feta of specific attributes can be qualified and consolidated on the market through networking and cooperation; and
3 how the encapsulation of the regional dairy cluster could be shifted from a sectoral approach (food chain) to a territorial perspective by integrating as many small traditional dairy territories in the Thessaly region as possible (human, natural, cultural milieu).

Activating territorial specificities under the Greek Feta PDO regime: the case of the Thessaly region

This chapter outlines the outcomes of the diagnostic phase and the elaboration of a territorial strategy to develop typical Thessalian dairy products through the Euro-Mediterranean LACTIMED Programme.[3] The main objective of the Programme was to reinforce cooperation among local actors involved in the value chain (inter-professional networks) and small dairy territories inside the region (cluster type) in order to activate latent territorial resources that contribute to place-based cheese qualities.

The sheep and goat dairy sector in Thessaly: an inherited pastoral territory in search of valorization of its specificities

The Thessaly region represents an important dairy basin in Greece, with high-density small ruminant livestock (14 per cent of the country's livestock capital) and a concentration of 75 dairy factories – mainly SMEs – that produce 30,300 tons of Feta (33 per cent of national production), among other products. About half of its surface area is fertile lowland surrounded by mountains, including the Pindus range and Mount Olympus.

The region has historically shaped traditional socio-productive systems based on extensive pastures and semi-nomadic pastoralist populations (mainly of Vlach origin) within the complementarity of natural resources and production systems in the mountains and the plain. This relationship profoundly changed after the 1960s, with the modernization of agriculture, the abolition of the set-aside system and the establishment of the productivist model in the plain (cereals and industrial crops, notably cotton), which pushed herds to hilly and mountainous areas.

The vast majority of cheese dairies are family-owned SMEs, with modern equipment and qualified owners, cheese makers and managers. The owners are often second or third generation proprietors of Vlach origin. They combine modern knowledge and traditional cheesemaking techniques, and are related to a network of cheese makers, usually of the same ethnocultural descent, inherited along with the family dairy. A handful of small territories have been identified within Thessaly, representing dairy basins that bring together small and medium-sized dairies, as well as their network of stockbreeders (see Figure 8.1). These small territories more or less coincide with the settlement areas of former semi-nomadic populations, granting them a significant tradition and reputation in stockbreeding. Cheese makers and stockbreeders are linked together by relations of trust and the feeling of belonging to a common social and cultural territory. However, there is no cooperation between the actors involved in the production chain within the same small territory, or even between small territories, in order to build common strategies for promoting local dairy heritage and a variety of typical cheeses. This reflects a severe functional introversion on the part of each of the actors.

The main dairy product in Thessaly is Feta. The region takes advantage of the growing demand for Greek Feta PDO and other internationally emerging dairy products such as yogurt and goat's cheese. However, more than ten years after the final listing as a Greek PDO product (2005), the price of Feta at the cheese factory remains stagnant while stockbreeders and cheese processors do not receive any significant added value from being involved in the PDO process. There are three reasons for this: (i) the national reach of the PDO label and thus the standardization of quality reflects an undifferentiated "national *terroir*"; (ii) weak promotion of the special characteristics and variety of flavours at the small territory level; and (iii) the absence of collective organization and cooperation of local actors to defend local specificities and negotiate vis-à-vis major dairy industries and extra-local actors (large distribution networks and retail).

The inability to overcome these issues and, especially, to increase the value of the product, is pushing large and medium-sized companies to increase their

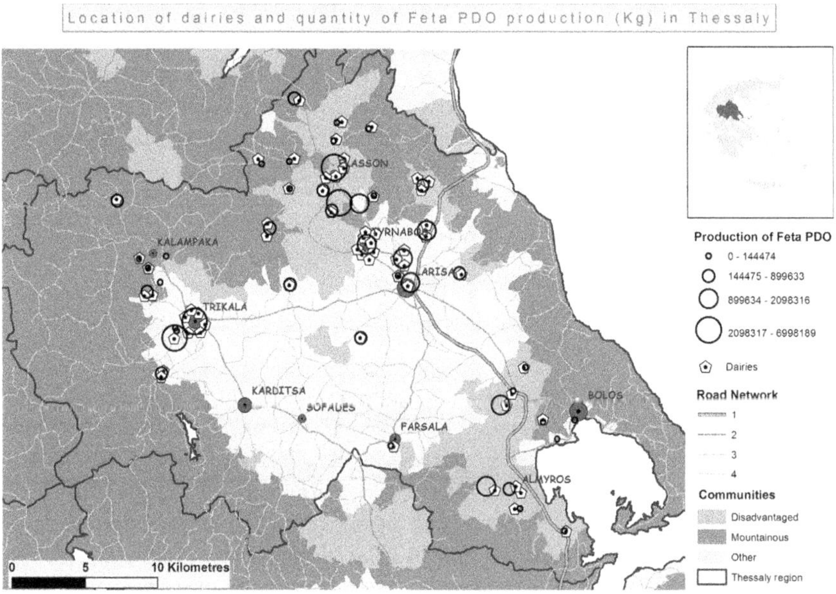

Figure 8.1 Distribution of cheese factories in the Thessaly region: an archipelago of small dairy territories.

Source: Laboratory of Rural Space, University of Thessaly.

cheese production. This strategy, in turn, leads them to seek milk outside the traditional area of collection (small territory of belonging), thus encouraging the creation of large, intensive livestock farms to the detriment of the organoleptic qualitative characteristics of Feta (industrial-style Feta). At the same time, it reinforces an asymmetric internal competition by means of economies of scale, which contributes to the marginalization of smaller dairies in isolated areas.

In contrast, artisanal family dairies operate in small territories in close cooperation with their networks of extensive stockbreeding farms, achieving relatively lower economic performance (small quantities of processed milk, underemployed facilities, etc.) but keeping high quality standards of the Feta product. These are both organoleptic (linked to *terroir*) and industrial (ISO – International Organization for Standardization and HACCP – Hazard Analysis and Critical Control Points standards) through the modernization of cheese factories, especially after the late 1990s with the support of the Common Agricultural Policy (CAP). These cheese processors conserve the quality of artisanal techniques and traditional practices, an advantage valued by the local/domestic markets which they mostly target. However, although small dairies are present in the regional and national markets, they have recently been seriously affected by the lack of cash liquidity caused by the economic crisis (delays in repayments from distribution networks and the retail market, suspension of bank lending). This weakness undermines the traditional long-term relationship with the stockbreeders of their territory with the risk of losing stable suppliers of raw material. It is these small artisanal units who are interested in cooperating in the context of a local networking of dairy SMEs and their small territories, in the search for alternative markets that recognize the local-differentiated-specific quality of Feta.

Elaborating methods and strategies for networking dairy actors: cooperation of Thessaly small dairy territories

Research context and problematic

The empirical research material was gathered through consultations with diverse stakeholders in the supply chain as well as semi-structured interviews with a sample of selected cheese processors. The results showed that *milk* is the key resource for the identification and qualification of the specific organoleptic attributes of Feta. Going back upstream to the genetic resources of the production chain (breeds adapted to local biophysical conditions), it is the grazing-based diet of small ruminants (grassland varieties in pastures) and traditional breeding practices (semi-extensive systems) that impart specific attributes and added value to the milk resource (Lambert-Derkimba *et al.*, 2011; Bele *et al.*, 2018). The activation of this currently undervalued potential would require collective processes of resource qualification, and the cooperation of local actors, inasmuch as local specificities in a given territory are shaped over time and reproduced by community-shared knowledge that constitutes a cultural patrimony (Muchnik *et al.*, 2008). The activation process involves all actors in the local pastoral community, especially breeders (animal feeding system) and cheese processors (selection of milk suppliers) who, through their choices, influence the quality of the final product. According to the existing literature, the establishment of an "origin-linked quality virtuous circle" as a methodology to support local stakeholders might allow the local community to identify, promote, preserve and reproduce a place-based specific quality of Thessalian Feta and the local resources embedded in it (Belletti *et al.*, 2012; Vandecandelaere *et al.*, 2009: 2–7).

However, as already mentioned, existing informal networks between dairy factories and associated breeders present a marked introversion; there is a complete lack of cooperation between cheese processors at the local (small territory) and regional (Thessaly region) level, despite the fact that there is geographical, socio-professional or ethnocultural proximity. The lack of interaction among territorial actors and the maintenance of strictly individual strategies is due to various factors, such as: (i) the family nature of dairies (it is the family that preserves the relationships with farmers and the pastoral community); (ii) the current market euphoria related to the increased demand for Feta PDO; and (iii) the organizational deficit at a regional level (no inter-professional organizations, dissolution of large cooperatives) combined with weak institutionalization within the environment where the public sector operates.

That said, only small and medium-sized artisanal processors of Thessaly were motivated to engage in a process of activation of dairy specificities in a sustainable virtuous quality circle. It is of concern for those who still maintain relationships with traditional breeders through networks of cultural belonging to a particular small territory (i.e. the aforementioned traditional dairy basins). In fact, seven small and medium-sized artisanal cheese factories, including a breeder cooperative (Shepherd Association of Olympus) have responded to such an initiative, proposed by the LACTIMED programme, and aim to jointly produce an origin-linked Thessalian Feta (territorial brand) within the national PDO scheme.

In this context, the main question to address was: How can cheese processors, the key actors in the process of product qualification and the promotion of a Thessalian quality Feta, be networked within the PDO regime? What they have in common is context-specific knowledge and socio-cultural ties which, however, are not shared through institutionalized professional structures; this leaves them powerless in the face of difficult conditions of market competition. That said, what drives this first group of determined cheese processors (small and medium-sized) – especially at this time of economic and financial crisis – is the awareness that they possess significant territorial assets which confer on them a qualitative advantage in answer to globalized market norms (large volumes of treated milk, economies of scale) that they cannot achieve. The participation of local actors in the territory's collective project represents a voluntary and solidarity commitment to coordinating their actions and promoting Thessalian Feta as a territorial brand name, highlighting a specific "local-based *terroir*" within the "national PDO *terroir*". However, as has been argued in the relevant literature, some involvement of external actors (government authorities, NGOs, research and development centres, etc.) is a necessary precondition in order to provide a supportive framework and empower small and medium-sized actors (Bowen, 2010: 209; Vandecandelaere *et al.*, 2009: 4). This role of external support in the Thessaly region was initially assumed by the EU LACTIMED Programme.

Elaborating territorial strategies

The strategy developed within the Programme aimed at enhancing the competitiveness of local dairy systems from both a sectoral and a territorial perspective by: (i) improving the production conditions of traditional extensive farming systems, thus contributing to the sustainability of territorial resources (e.g. locally adapted breeds, native flora); and (ii) highlighting the place-based distinctiveness of milk as the key territorial resource that confers specific quality and flavours on Feta cheese (see Figure 8.2). For this purpose, an expert team from the University of Thessaly and fellow research laboratories

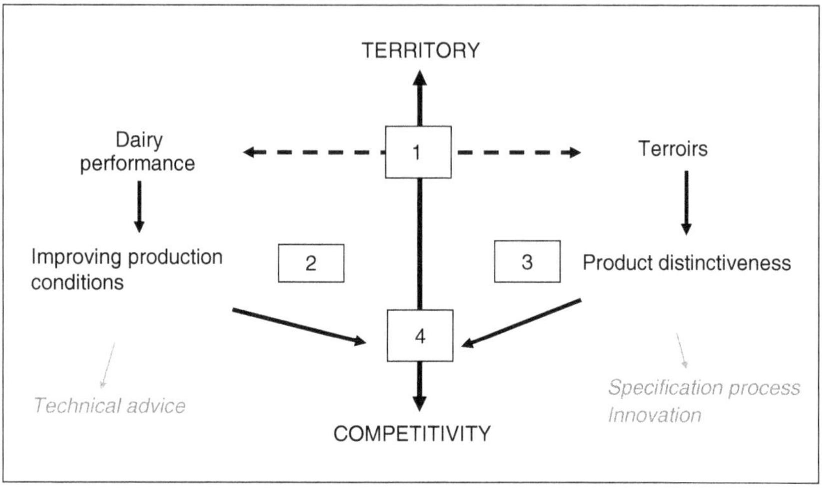

Figure 8.2 Enhancing competitiveness through sectoral and territorial perspective.
Source: Laboratory of Rural Space, University of Thessaly.

was organized to provide scientific and technical support to the territorial collective project. More specifically, the team provided extension services in improving the forage quality of pastures and applying best practices in grazing management; increasing ewe milk production through a balanced animal diet; improving the quality of the milk, through a diet based on pastures' native plant species, combined with eco-friendly local fodder crops; and a locally produced diet that also leads to a significant reduction in production costs and ecological footprint in comparison with dominant practices (proximity agriculture).

Discussions within the body of the first seven cheese processors and associated actors in the production chain (their milk suppliers, fodder farmers) allowed us to define the specific characteristics of the products of small territories and the production process that confer particular organoleptic properties and taste quality on Thessalian Feta. The qualification phase resulted in a code of practice elaborated within the "origin-linked quality virtuous circle", according to which the Thessaly Feta label is manufactured under specific conditions (beyond the national PDO requirements). These are: (i) the cheese is made from the milk of locally adapted breeds of sheep and goats that graze in the area's pastures; (ii) flocks feed on supplementary fodder based on local farming crops; (iii) the necessary period for cheese ripening is three months; (iv) cheese makers use artisanal production and ripening methods (e.g. ripening in barrels); and (v) cheese processors should be of a small or medium size to ensure artisanal methods and the integration of locally anchored knowledge and know-how. Internal control of quality at all stages of the production chain will be assured by means of a participatory guarantee system, which would contribute to a sense of cooperation and trust among the actors involved. Moreover, the integration of the principles of reciprocity and fair redistribution of added value upstream of the production chain enhance transparency and loyal cooperation between local stakeholders.

Implementation of cooperative procedures and territorial governance

The organizational deficit in the construction and valorization of specific resources at the regional level was compensated through awareness and mobilization of all involved actors along the value chain, combined with successive occasions of rapprochement (targeted informative and consultation meetings, focus groups, etc.) from the perspective of social interaction, reconciliation and adaptation to an increasingly competitive economic environment. Although socio-cultural intimacy may be consistent with geographical proximity within small dairy territories, thus shaping informal networks of collaboration (verbal agreements) – such as Vlach cheese makers supplying milk through a local intergenerational network of Vlach stockbreeders – this associational factor disappears at a larger regional scale (Thessaly region). Wider partnerships, notably among cheese factories, are thus consolidated. Yet, stiff competition between dairy producers due to the increasing dependence of the regional economy on global market dynamics intensifies the risk of small dairies disappearing (Enriquez-Sanchez *et al.*, 2017). In this context, the interaction within the successive occasions of rapprochement between different stakeholders (dairy actors, local development agencies, commercial and industrial chambers, cooperative banks, municipal services, etc.) on a wider regional level might create reciprocity and a spirit of cooperation, and therefore restore the territorial anchorage of collective knowledge and practices. This process also brings local actors closer to the service sector and helps them become familiarized with the institutional environment of the public sector and existing policies (Dalle and Didry, 1998: 312).

The signing of a Charter of Cooperation is a first step of institutional recognition of informal agreements among regional stakeholders that constitute the Territorial Assembly (see Figure 8.3). The Assembly is a broad-based representation of actors who, in one way or another, are involved in the value chain, including both those that are directly involved in the dairy supply chain and come from small dairy territories (i.e. producers and processors) and others that operate outside it (regional research institutions, local development agencies, cooperative banks, municipalities as pasture managers, etc.). The mission of the Territorial Assembly is to ensure the strategy and goals of the project established by the Charter (Angeon *et al.*, 2006) as a means of territorial governance are implemented. The Charter reflects the cooperation of the small dairy territories of Thessaly in a cluster form representing "a kind of Archipelago of places or micro-basins" within the national *terroir* of Feta PDO (Requier-Desjardins, 2010; Sforzi and Mancini, 2012).

Territorial coordination: linking dairy territories, local actors and consumers

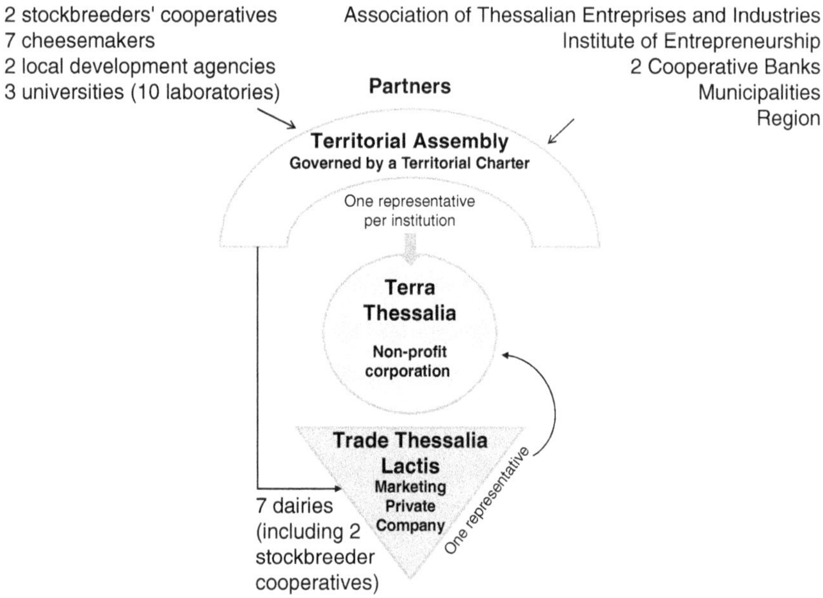

Figure 8.3 Cooperation of small dairy territories in the region of Thessaly.

Source: Laboratory of Rural Space, University of Thessaly.

A smaller circle including actors directly involved in and/or supporting the production process (i.e. cheese processors and the stockbreeders gathered around them, as well as financial, scientific and technical support providers such as research laboratories, cooperative banks and local development agencies) represents the central core of the territorial project under the name "Terra Thessalia" in the form of a non-profit corporation. This label also constitutes the territorial brand name of Thessaly Feta. The Terra Thessalia corporation – a limited liability company of low investing/start-up capital (Trade Thessalia Lactis) set up by seven cheesemakers including breeder cooperatives engaged with the project – has undertaken the marketing of dairy products ensuring fair redistribution of trade gains among producers (notably stockbreeders), a supportive advisory structure and micro-investments. Terra Thessalia, a conceptualization that is both intangible (regional brand name and reputation) as well as tangible (a business structure), is the key activator towards democratic territorial governance of the dairy chain as well as territorial development at regional level. In this context, the Terra Thessalia corporation is not considered to be just a typical economic cluster (agglomeration economies) but a territorial cluster mobilizing collective resources and cultural heritage for social benefit (local communities, consumer awareness, fair trade and solidarity economy networks).

Supporting specific dairy resources and territorial cooperative action: setting up a territorial participatory quality guarantee system

The basic idea was to design a participatory guarantee system to make the distinctive quality of Thessaly dairy products visible to consumers, as well as boost their marketability at a fair price for all actors involved in the food supply chain, including small producers in remote areas. The main objective was to create a methodological tool to guarantee quality claims related to place-based attributes and establish a direct relationship between producers and consumers through information and traceability systems. For this purpose, the Terra Thessalia corporation implemented a Territorial Participatory Guarantee System (TPGS), the aim of which was to reveal the specific characteristics of the resource (milk being the key resource) of different small dairy territories (dairy *terroirs*) within the national PDO quality definition (Cuéllar-Padilla and Ganuza-Fernandez, 2018). The TPGS relied on elaborate informative and technical material provided by experts on production systems, techniques and qualification benchmarks that must be met for this particular purpose (See Table 8.1).

The TPGS was conceived along two dimensions to give evidence on: (i) the traceability of milk production procedures and cheese making in

Table 8.1 Quality claims and guarantee means of the Territorial Participatory Guarantee System

What	*How*
• Animal welfare and feeding quality (grazing practices)	• Movement of herds using Global Positioning System (GPS) • Analysis of pasture flora
• Contribution to reducing environmental footprint	• Proximity between stable, dairy and livestock feed production areas
• Indigenous breeds (strength/resistance), animal hygiene	• Official public service documents, identification, monitoring by Terra Thessalia experts (epidemics, antibiotics, etc.)
• Optimization of the milk collection period (quality of the flora)	• ISO system etc. • Milk transport control
• Duration of ripening	• Batch codes/ISO
• Territorial quality mark (sustainable management of natural and cultural heritage resources, Natura 2000)	• Evaluation based on spatial, environmental, etc. criteria
• Product quality: milk and cheese characteristics (local-based feed, herd movement, etc.)	• Analysis performed by competent laboratories

Source: Laboratory of Rural Space, University of Thessaly.

accordance with the code of practice, i.e. integration of selected flocks in a particular semi-extensive stockbreeding system to ensure the specific quality claims under the Terra Thessalia label (brand name); and (ii) certification of the livestock farms and flocks participating in the territorial project of Thessalian Feta. The TPGS represents both an informative tool that guarantees the authenticity of the product (links with the natural and cultural milieu, nutritive and health qualities, etc.) and a territorial strategy to (re)integrate livestock farms and artisanal cheese factories in their small dairy territory (*terroir*). It thus brings in the additional guarantees and transparency of third-party certification through a private territorial label (Terra Thessalia brand name). That said, the TPGS strengthens the traditional dairy sector within small territories without requiring additional (costly) legislative certification.

The TPGS was also conceived to be open to the integration of new cheesemakers and stockbreeders willing to share Terra Thessalia quality principles, thus enhancing the legitimacy of the system and the consolidation of the Terra Thessalia label through territorial cooperation. Those wishing to adopt the TPGS must meet the following criteria: agropastoral practices; native breeds; artisanal cheesemaking techniques; and traditional dairy products. This means they respect the specifications of both the PDO and Terra Thessalia.

The multiple roles of the TPGS are decisive in the adoption of the territorial approach to governance and rural development at regional/local level. The TPGS is a support tool for the management of local natural and cultural resources, but it also reinforces participative democracy while avoiding conflicts due to the use of new technologies (GIS, virtual representation, ICT, etc.). After the completion of the Terra Thessalia label procedures and the start-up of the trading company of cheesemakers, the first tentative market sales abroad (niche markets, especially in Switzerland and France) and in the domestic market (short supply chains and solidarity markets in Athens) have confirmed consumers' interest in non-industrial place-based Feta cheese. The recognition of the superior quality of Thessalian Feta confers an important price premium in comparison with conventional Feta of industrial quality. This reflects the acknowledgement of a heritage activity (semi-extensive farming) which is undervalued by the quantitative approach of intensive production systems and the norms of the industrial model that prevails in the global market.

Conclusion

As has been highlighted by scholars (Bowen, 2010; Hinrichs, 2003; DuPuis and Goodman, 2005), internationally renowned GI products are often threatened with losing their embeddedness by extra-local actors that tend to appropriate niche markets so as to maximize their commercial profit and yet do not contribute to sustainable production systems. We argued that the establishment of a place-based label for Feta (Terra Thessalia brand name) within the defined PDO area by activating territorial assets (mountainous pastures, locally adapted breeds, cheesemaking techniques and traditions, ethnocultural proximity, etc.) fosters small dairy territories – i.e. traditionally specialized and historically constructed territories that are firmly embedded in ecological, social and cultural networks. To enhance consumer loyalty within a virtuous circle of origin-linked quality, a Territorial Participatory Guarantee System has been created to assure the authenticity of the product (physical and cultural ties with the Thessalian *terroir*) compared to Greek Feta PDO of standard quality. Moreover, Terra Thessalia Feta supports artisanal dairies and actors involved in the supply chain (stockbreeders, fodder producers, local food processors, traders, local development agencies, municipalities acting as managers of communal pastures) by activating common knowledge and socio-cultural ties.

The cooperation of small dairy territories through a territorial brand (the Terra Thessalia label) inside the "national *terroir*" (nation-wide PDO) and, furthermore, the cooperation of a dynamic core of artisanal dairies representing this "Archipelago of micro-basins" enables participatory territorial governance of the dairy food chain and social control of local cultural heritage and collective intellectual rights vis-à-vis major external actors. This perspective can thus facilitate the sustainability of the localized agri-food system in the context of a tight oligopolistic food market in Greece.

Notes

1 This chapter is based on further elaboration of the unpublished paper prepared by the authors for the one hundred and forty-fifth European Association of Agricultural Economists seminar "Intellectual Property Rights for Geographical Indications: What is at Stake in the TTIP?", Parma, 14–15 April 2015 (Anthopoulou and Goussios, 2015).

2 The concept of "localized agri-food systems" (SYAL in French or LAFS in English) was conceived in the mid-1990s by the French Agricultural Research Centre for International Development (CIRAD). Since then it has been developed both as a research topic and a territorial development strategy. It aims at developing a specific theoretical framework for understanding the organization and operation of a set of economic (production, processing, catering, etc.) cultural (education, festivities, etc.) and recreational (rural tourism, training, etc.) activities interwoven in the same territory, from both a material and a symbolic point of view (Cañada and Muchnik, 2011; Bowen and Mutersbaugh, 2014). This is in order to understand the synergies between various territorial activities so as to reinforce the anchorage and develop the identity of typical GI products (Muchnik *et al.*, 2008).

3 European funded European Neighbourhood and Partnership Instrument – Cross Border Cooperation in the Mediterranean Programme, "Promoting Mediterranean Dairy Products" (www.lactimed.eu/en).

References

Angeon, V., Caron, P. and Lardon, S. (2006). Des Liens Sociaux à la Construction d'un Développement Territorial Durable: Quel Rôle de la Proximité dans Ce Processus? *Développement Durable et Territoires*, [online] Dossier 7, 23 pages. Available at: http://developpementdurable.revues.org/2851 [accessed 28 March 2018].

Anthopoulou, T. and Goussios, D. (2015). Activating territorial specificities under a national PDO cheese label. Cooperation of small dairy territories to promote local-placed Feta in Thessaly Region. In: Proceedings of the 145th EAAE Seminar, *Intellectual Property Rights for Geographical Indications: What is at Stake in the TTIP?*, 14–15 April 2015, University of Parma.

Bele, B., Norderhaug, A. and Sickel, H. (2018). Localized Agri-Food Systems and Biodiversity. *Agriculture*, [online] 8(2), 22. Available at: doi.org/10.3390/agriculture8020022 (accessed 15 May 2018).

Belletti, G., Casabianca, F. and Marescotti, A. (2012). Local Food Quality and Local Resources. In: F. Arfini, M.C. Mancini and M. Donati (eds.), *Local Agri-Food Systems in a Global World: Market, Social and Environmental Challenges*. Newcastle upon Tyne, UK: Cambridge Scholars Publishing, pp. 71–95.

Bowen, S. (2010). Embedding Local Places in Global Spaces: Geographical Indications as a Territorial Development Strategy. *Rural Sociology*, 75(2), 209–243.

Bowen, S. and Mutersbaugh, T. (2014). Local or Localized? Exploring the Contribution of Franco-Mediterranean Agrifood Theory to Alternative Food Research. *Agricultural Human Values*, 31, 201–213.

Campagne, P. and Pecqueur, B. (2014). *Le Développement Territorial. Une Réponse Émergente à la Mondialisation.* [online] Paris: Charles Léopold Mayer. Available at: www.eclm.fr/ouvrage-368.html [accessed 15 May 2018].

Cañada, J. and Muchnik, J. (2011). Introduction : Ancrage et Identité Territoriale des Systèmes Agroalimentaires Localisés. *Economie Rurale*, [online] 322, 4–10. Available at: 10.4000/economierurale.2962 [accessed 15 May 2018].

Cuéllar-Padilla, M. and Ganuza-Fernandez, E. (2018). We Don't Want to Be Officially Certified! Reasons and Implications of the Participatory Guarantee Systems. *Sustainability*, [online] 10(4), 1142. Available at: doi.org/10.3390/su10041142 [accessed 15 May 2018].

Dalle, J.M. and Didry, C. (1998). Les Approximations de la Proximité Comme Catégorie de L'action Publique. In: M. Bellet, T. Kiriat and C. Largeron (eds.), *Approches Multiformes de la Proximité*. Paris: Hermès, pp. 309–327.

DuPuis, E.M. and Goodman, D. (2005). Should We Go "Home" to Eat? Toward a Reflexive Politics of Localism. *Journal of Rural Studies*, 21, 359–371.

Enriquez-Sanchez, J., Munoz-Rodriguez, M., Altamirano-Cardenas, J.R. and Villegas-de Gante, A. (2017). Activation Process Analysis of the Localized Agri-Food System Using Social Networks. *Agricultural Economics [Zemědělská Ekonomika]*, 63(3), 121–135.

Hinrichs, C. (2003). The Practice and Politics of Food System Localization. *Journal of Rural Studies*, 19, 33–45.

Lambert-Derkimba, A., Verrier, E. and Casabianca, F. (2011). Tensions entre Ressources Génétiques Locales et Ancrage Territorial des Produits. La Race Porcine Corse dans un Projet AOP. *Economie Rurale*, 322, 39–49.

Marescotti, A. and Belletti, J. (2016). Differentiation Strategies in Coffee Global Value Chains through Reference to Territorial Origin in Latin American Countries. *Culture & History Digital Journal*, [online] 5(1). Available at: doi.org/10.3989/chdj.2016.007 [accessed 15 May 2018].

Muchnik, J. (2006). Nourrir le Corps Humain et le Corps Social. In: B. Hubert and O. Clément (eds.), *Le Monde Peut-il Nourrir Tout le Monde?* Paris: Quae Editions, pp. 25–42.

Muchnik, J. and Sautier, D. (1998). *Systèmes Agro-Alimentaires Localisés et Construction de Territoires*. Paris: ATP CIRAD.

Muchnik, J., Canada, J. and Salcido, G.T. (2008). Systèmes Agroalimentaires Localisés: État des Recherches et Perspectives. *Cahiers Agricultures*, [online] 17(6), 513–519. Available at: doi.org/10.1684/agr.2008.0251 [accessed 10 February 2018].

Muchnik, J., Requier-Desjardins, D., Sautier, D. and Touzard, J.M. (2007). Les Systèmes Agroalimentaires Localisés. *Economie et Société* (Série Systèmes Alimentaires), 29, 1465–1484.

Requier-Desjardins, D. (2010). L'évolution du Débat sur les SYAL: Le Regard d'un Économiste. *Revue d'Economie Régionale & Urbaine*, [online] 4, 651–668. Available at: doi.10.3917/reru.104.0651 [accessed 9 September 2018].

Rytkönen, P., Bonow, M., Girard, C. and Tunón, H. (2018). Bringing the Consumer Back in – The Motives, Perceptions, and Values behind Consumers and Rural Tourists' Decision to Buy Local and Localized Artisan Food – A Swedish Example. *Agriculture*, [online] 8(4), 58. Available at: doi.org/10.3390/agriculture8040058 [accessed 15 May 2018].

Sforzi, F. and Mancini, M.C. (2012). A Reinterpretation of the Agri-Food System and Its Spatial Dynamics through the Industrial District. In: F. Arfini, M.C. Mancini and M. Donati (eds.), *Local Agri-Food Systems in a Global World: Market, Social and Environmental Challenges*. Newcastle upon Tyne, UK: Cambridge Scholars Publishing, pp. 9–28.

Torres Salcido, G. and Muchnik, J. (2012). Globalisation/Fragmentation Process: Governance and Public Policies for Localized Agri-Food systems. In: F. Arfini, M.C. Mancini and M. Donati (eds.), *Local Agri-Food Systems in a Global World: Market, Social and Environmental Challenges*. Newcastle upon Tyne, UK: Cambridge Scholars Publishing, pp. 97–116.

Vandecandelaere, E., Arfini, F., Beletti, G. and Marescotti, A. (eds.) (2009). *Linking People, Places and Products: A Guide for Promoting Quality Linked to Geographical Origin and Sustainable Geographical Indications*. Rome: FAO-SINER-GI. Available at: www.fao.org/docrep/013/i1760e/i1760e00.pdf [accessed 9 September 2018].

Part III

Alternative agri-food market channels

This section examines short food supply chains (SFSCs) as efforts to provide alternative market channels in relation to global food supply chains, which are typically organized on the basis of market channels provided by transnational corporations. The chapters in this section highlight that developing and being engaged in SFSC initiatives is an important political activity as these initiatives provide alternative production and consumption models to their capitalist counterparts. Yet, this political dimension is often underscored as the political values projected by SFSC initiatives are not necessarily widely shared in society. Accordingly, the chapters in this section examine whether SFSC initiatives are likely to remain a market niche, as in community supported agriculture (CSA), or whether they have the potential to become more mainstream, as in the case of alternative food retailers. However, by trying to become mainstream do SFSCs become less "alternative"?

9 The fairness of alternative food retailers

Agni Kalfagianni

Introduction

Questions of fairness and justice become increasingly important in a world with growing inequalities within and beyond state borders (Steffen *et al.*, 2015; Wettstein, 2009), and are increasingly pertinent in the agri-food system. Intensified agriculture has been recognized as a key driver behind many environmental threats, including climate change, biodiversity loss and degradation of land, and is pushing the environment beyond "planetary boundaries" (Foley *et al.*, 2011). Yet, at the same time, almost 870 million people are chronically undernourished, with the number of hungry people remaining "unacceptably high" according to the Food and Agriculture Organization (FAO, 2012: 8). In the poorest parts of the world, especially Sub-Saharan Africa, one in four people are chronically hungry (FAO, 2014) while two-thirds of the population work in agriculture (FAO, 2008). Ironically, one-third of all food production is wasted throughout the supply chain[1] while the amount of food wasted in Europe and North America equals the total net food production of Sub-Saharan Africa (FAO, 2011).

In these circumstances demands for fairness and justice are expected not only to become particularly fervent but are likely to determine the effectiveness and legitimacy of the overall agri-food governance architecture.[2] Indeed, the subject of justice in agri-food governance can no longer be regarded as a purely academic and philosophical inquiry but must now be considered a fundamental political concern (see Forman and Mackie, 2013).

A key actor with the potential to address the multiple dimensions of (in)justice that are associated with the global agri-food system is alternative food retailers, i.e. retailers that prioritize food that is prepared or produced by non-conventional agricultural systems. In contrast to conventional retail corporations, this type of retailer offers food (and other) products that have almost exclusively been produced in some "sustainable" way. These include, for instance, eco and Fairtrade certified products and food produced locally. Due to their position in the supply chain, alternative food retailers offer a key outlet to sustainable-minded producers and consumers. Surprisingly, the literature on short food supply chains has, so far, paid little attention to alternative food retailers.

This chapter examines the role of alternative food retailers in fostering a more sustainable agri-food system, paying particular attention to fairness and justice. The chapter first conceptualizes justice in the agri-food system as a multidimensional concept consisting of material, social, environmental and political dimensions. In doing so the chapter highlights key debates regarding the extent to which conventional and short food supply chains address these justice dimensions. The chapter subsequently evaluates discursively the extent to which four alternative food retailers in the Netherlands recognize the four justice dimensions in their practices. It concludes by discussing the implications for a just transformation of the agri-food system.

Fairness and justice in the agri-food system

There is no common understanding in the literature regarding what constitutes justice, in general, and justice in the agri-food system more specifically. In attempting to conceptualize justice, an important strand of the literature focuses on social movements and their justice demands around food. In their work on the evolution of the US and global food system and the parallel development of the food movement, Gottlieb and Joshi (2010) define food justice as "ensuring that the benefits and risks of where, what and how food is grown and produced, transformed and distributed, and accessed and eaten are shared fairly" (p. 6). Other scholarship focuses specifically on race and class in the food system. This work underlines the limited ability of low-income communities and communities of colour to access – that is produce and consume – healthy food (Alkon and Agyeman, 2011). Still others argue that justice is not only about access, distribution and rights, but also about the capacity to provide critique to the conditions of production and visualize how these can be transcended (Besky, 2014).

The Fairtrade scholarship, in turn, examines justice though the impact on those involved, such as producers, artisans and their organizations (Le Mare, 2008). Justice is defined through the demands of its advocates and the values of the Fairtrade movement in promoting transparent and reciprocal relationships between producers and distant consumers and fostering democratic and gender inclusive ways of organizing within producers' organizations (Lyon and Moberg, 2010). Critical perspectives debate whether Fairtrade represents roll-out neoliberalism or a Polanyian counter movement (Bacon, 2010; Guthman, 2007). In this context, justice is broadly understood as challenging the dynamics of capitalism which is believed to be the main cause of injustice in the agri-food system (Guthman, 2007).

Drawing on this literature and previous work on this issue (Kalfagianni, 2014, 2015), this chapter argues that justice in the agri-food system involves not only material demands for access and distribution but also recognition of cultural differences and the need for political emancipation. Further, the chapter underlines that a rather neglected but equally important dimension of justice in the agri-food system is responsibility towards non-human species and future generations. This results in a multidimensional concept that includes the material, social/cultural, environmental and political dimensions of justice.

Material dimension

The material dimension of justice, in the context of the agri-food system, refers primarily to the ability to access sufficient, safe and nutritious food, i.e. achieve food security.

Conventional food supply chains negatively affect food security in numerous ways. First, scholars note the uneven trade and investment relationships prevalent in global supply chains which incur myriads of environmental and social costs that tend to fall disproportionally on the world's poor, thus indirectly affecting their food security (Dauvergne, 2008). Second, conventional supply chains have been associated with the consolidation of the retail end and the associated ability of large retail corporations to control prices. Reports reveal that many supermarkets are selling food products at prices which outstrip the cost of production manyfold, without corresponding benefits for farmers (European Parliament, 2009; Harward, 2010; OECD, 2010; Thanassoulis, 2008). Third, scholars note the increasing financialization of the conventional food system, the "increasing importance of financial markets, financial motives, financial institutions, and financial elites in the operation of the economy and its governing institutions, both at the national and international levels" (Epstein, 2005:3). Financialization has been associated with speculation in agricultural commodity futures markets resulting in price volatilities and land grabbing effects (Clapp, 2014).

The extent to which short food supply chain initiatives can address these concerns is debatable, however. Some scholars note the opportunity to increase access and food security as a result of directly participating in the production of food (Gottlieb and Joshi, 2010). Studies suggest that producers enjoy a price premium (Pearson *et al.*, 2011), that the elimination of the "middleman" enables farmers to receive a greater share of the profits (Sage, 2003) and that short food supply chain initiatives provide growers with an opportunity to diversify and add value to produce that would not usually be marketed (Alonso, 2011). In contrast, others highlight the inability of short food supply chains to address the diversity in educational and occupational backgrounds (Hinrichs and Kremer, 2002), their disproportionate attention to farmers in relation to other vulnerable actors, such as farm labour, and their inaccessibility to low-income consumers (McEntee, 2011).

Social/cultural dimension

The social and cultural dimension of justice includes questions of recognition of labour rights, race, gender and cultural diversity.

Agriculture is a highly unregulated field in many countries with adverse consequences for farm labour. The FAO estimates that nearly 3.5 million people suffer from forced labour in agriculture; only 10 per cent of rural workers are unionized; about 60 per cent of the world's child labour is involved in agriculture; and 43 per cent of the total agricultural workforce consists of women whose working life is constrained by various biases and discriminations.[3] In developing countries, in particular, farming practices are characterized by an increasing feminization, with women accounting for 60–80 per cent of food production (Agarwal, 2014; Angotti, 2012; Deda and Rubian, 2004; Fuchs *et al.*, 2012; Singh, 2003). Furthermore, a large number of migrants originate from rural areas and about 40 per cent of international remittances that workers send to their home countries go to rural areas (FAO, 2016). With regard to migrant agricultural labour, more specifically, studies report the "invisibility" of migrant farmworkers in large prosperous areas of the Global North, associated with the practices of marginalization, oppression and racism (Brown and Getz, 2011).

There are several examples indicating that short food supply chains can address these concerns, but no systematic evidence in this context. In the US, for instance, studies report how Latino community gardeners cultivate staples essential for their cultural identity (Alkon and Agyeman, 2011). Similarly, communities of colour create their own cultural spaces that reflect their heritage purposefully, so differentiating themselves from "whites" (ibid.). Likewise, research highlights the occupation of lost farmland due to urban sprawl by migrant and refugee farmers and their efforts to build a community, thus creating a "new type of farmer" (Gottlieb and Joshi, 2010: 134–138). However, scholars also warn against romanticizing an alternative food utopia around the "local" and advise, instead, adopting a reflexive attitude that is not based on fixed norms and imaginaries of perfection (that can also be later coopted by corporate actors) (Goodman *et al.*, 2012).

Environmental dimension

Justice in the environmental dimension is concerned with the disproportionate negative environmental impacts on the poor, and also more broadly about human ethical obligations to non-human species and future generations.

Agriculture occupies about 38 per cent of the earth's terrestrial surface (Foley *et al.*, 2011) and uses 70 per cent of globally extracted water.[4] Only 62 per cent of global crop production is dedicated to direct human consumption, with 35 per cent used for animal feed and 3 per cent for bioenergy, seed and other industrial products (Foley *et al.*, 2011). In Europe and North America 40 per cent of crops are devoted to human consumption while in Africa and Asia the figure is 80 per cent. At the same time, food production is responsible for 25 per cent of total greenhouse gas (GHG) emissions (Tilman and Clark, 2014), as well as marine and fresh water pollution and loss of habitat and biodiversity (Runhaar *et al.*, 2016). Furthermore, as noted earlier, about one-third of total food production is wasted worldwide. In the EU-28, food waste (including edible food and non-edible parts associated with food) was

173 kg per person in 2012, which amounts to 20 per cent of total food produced (Stenmarck *et al.*, 2016). In the US the amounts are larger and food waste constitutes about 40 per cent of food produced (Gunders, 2012). While in developing countries the figures are also high, the difference is that food waste occurs much earlier in the supply chain and not at the processing and consumption end. For instance, food waste per capita in Sub-Saharan Africa and south and south-east Asia is about 6–11 kg per year.[5]

In this context, scholars argue that short food supply chains tend to produce food in an environmentally sustainable way (Horrigan *et al.*, 2002). Farms participating in such chains tend to generate a greater variety of vegetables, fruits, grains and animal products compared to large-scale farms that are mostly subjected to monocultures (Altieri, 2009; Arias *et al.*, 2013). These diversified traditional crop-rotating systems are also considered more able to adapt to climatic differentiations, diseases and other natural struggles (Tsegaye, 1997). In addition, compared to mechanized agriculture, these systems have limited off-farm inputs and make less intensive use of resources. However, it is not yet clear who benefits from short food supply chains and whether they address environmental injustices, particularly for the poor.

Political dimension

The political dimension of justice highlights the aspect of agency, the ability to substantively participate in, critically reflect upon and/or set one's own rules in the governance of food. Agency is an important dimension of justice emphasized by both liberal and critical perspectives (Fraser, 2000; Nussbaum, 2000; Rawls, 1971; Sen, 1999). It emphasizes people's will to be active in their struggle even at the expense of other values, such as personal satisfaction. Agency, then, is "the unlimited self-questioning about the law and its foundations as well as the capacity, in light of this interrogation, to make, to do and to institute" (Castoriadis, 1991:86). It is only in the context of agency that the question "What is just?" has meaning.

Reflecting upon the conventional food system, scholars note that smallholder agency is particularly impaired by distancing; the geographical expanse of commodities along global supply chains and the knowledge gaps in the social and environmental impacts of food production affects the ability to influence the governance of food (Clapp, 2014). In contrast, the agency of corporate actors has increased accordingly. In this context, studies argue that short food supply chains may offer autonomy from corporate actors by allowing smallholders to source inputs locally or via direct access to the market (Santini *et al.*, 2013). Likewise, advocates of "food democracy" and "food citizenship" emphasize the opportunities provided by short food supply chains to actively participate in the construction of an alternative food imaginary (Seyfang, 2006; Wittman, 2009). Despite this promise, scholars also issue caution against the fetishization of the local as the prime place for agency in food governance, pointing to the appropriation of the language of democracy and locality by large retail corporations (Johnston *et al.*, 2009). Similarly, they warn against the binary distinction between the local and the global which may foster a "false" dilemma about which level of action it is worth being political about (Lockie, 2009). Instead, agency should recognize the multiplicity of avenues available for action and realize that no single "alternative" model can be set for the entire food production system (ibid.).

An alternative retail model for fair food?

The study that follows focuses on four alternative food retailers in the Netherlands, representing about 30 per cent of the 448 alternative food stores all over the country, but about 78 per cent of stores with a supply chain organization (171 stores in 2015) (Detailhandel, 2016): Ekoplaza, Marqt, Natuurwinkel and Estafette. All four use a combination of proximity relations and extended relations. Ekoplaza consists of 74 shops, which makes it the largest retailer of its kind. Marqt, although it currently has only 15 shops, was the first retailer that successfully combined fashionable food with that which is also sustainable (Keuning, 2015). Natuurwinkel (28 shops) and Estafette (18 shops) also represent significant alternative food retailers in the Netherlands (Natuurwinkel, 2016; Estafette, 2016).

Ekoplaza

The organic food retailer Ekoplaza is owned by parent company Udea and was founded in its current form in 2010. Ekoplaza is the biggest alternative food retailer in the Netherlands with 74 shops all over the country. It aims to provide organic food "for every consumer" both in shops and online. Udea is a family owned business. Apart from being the franchisor of Ekoplaza, a wholesaler, a house of brands and importer and exporter of food, Udea has also developed its own brands such as Proef, BioCumulus and 3 Little Goats (Udea, 2013a).

Marqt

Marqt is a food retail company founded in 2007 by Quirijn Bolle. At the time of writing, it has 13 shops in the cities of Amsterdam, Haarlem, Rotterdam, The Hague and Utrecht (NVM, 2015). The company has positioned itself within the food retailers' market with the slogan "real food", which means retailing food according to a set of predetermined principles: respect for nature, animals and human beings. The retailer's preference is to purchase products from local farmers and those who produce food that remains close to its origins, meaning no unnecessary changes or additives (Marqt, 2016). Marqt's main investor is Verlinvest, a family-owned investment holding company which has committed to providing enough capital for Marqt to expand to 20 shops (Smit, 2012).

Natuurwinkel

Natuurwinkel is part of the wholesale company Natudis. Natuurwinkel currently has 28 shops spread all over the Netherlands. The company was founded in Ermelo in 1981 with a vision that food could be better and healthier. Since its establishment, Natudis is concerned with the distribution of healthy food across all the Benelux countries. Natudis believes that a healthy planet is the basis for all life, and therefore also for our food. For this reason, it promotes itself as "the wholesaler and retailer of nature". Natudis has developed two own brands: Ekoland and LunaeTerra. The company itself manages the whole process of development, purchase and marketing (Natudis, 2016). Since 2014, the holding Vroegop Ruhe & Co. BV has owned Natudis. Apart from Natudis this holding consists of Vroegop-Windig, De Kweker and Kroon Biologische Verswaren.

Estafette

Estafette has 18 retail shops which are part of the Estafette Odin cooperative, a partnership between producers, wholesalers, retailers and consumers. All partners commit to the distribution of organic food. Estafette Odin originates from the company Odin, founded in 1983 on the basis of three principles: healthy agriculture with respect and care for the earth; the supply of high quality food with respect for people's health; and healthy economic practices with respect for people, resources and the earth (Estafette Odin, 2016). Estafette Odin adopted its cooperative structure in 2012 (Estafette Odin, 2016).

Material dimension

Food security is not explicitly addressed by any of the alternative food retailers examined here. Rather, what appears to be a prominent issue in the material dimension is fair prices for farmers, the authenticity of food and a distinction from corporatism.

As an illustration of this, Ekoplaza notes that money is not the main driver of its business but the belief that organic agriculture is better "for us and for the world". It emphasizes that the constant pressure on prices, a typical practice by conventional retailers, is at the expense of quality. Instead it is quality that should be determined first and the right price for that quality second.[6]

Likewise, Marqt highlights its differences with conventional retailers for whom low price and high volume is the main approach, resulting in low prices for farmers on the basis of which they cannot live. Thus, Marqt advocates fair prices for farmers by buying nothing from them at less than the cost price. Like Ekoplaza, Marqt argues that the reduction of costs is at the expense of farmers, working conditions for labour and the environment. It further emphasizes that, in the end, the price is paid by the consumer through worsening health, including allergies, chronic heart disease and obesity. In this context, Marqt highlights that it is not in its interest to become super-rich; this is only possible through corporate practices which have led to the huge inequalities the food system is currently experiencing.[7] Furthermore, addressing the criticism about the exclusionary character of alternative food retailers, particularly in relation to low income groups, Marqt emphasizes that consumers can afford higher prices if they change their purchasing behaviour: less meat, less food that they will throw away, etc. Changing consumption patterns will afford the world more food, not industrialized agriculture, according to Marqt (OneWorld, 2015).

Natuurwinkel makes no reference to prices and rather focuses on the environmental, health and animal welfare aspects of seasonal, organic and bio-dynamic food production.

Estaffete highlights the fairness of prices so that all involved in the supply chain can get what they have worked for but also so that consumers do not pay too much. It is not about personal profit but profit for the community. This can be achieved through a cooperative structure where members share their property and profits. This form of entrepreneurship also fits the present time, according to Estaffette, because consumers and employees are responsible for the activities of the company and all decisions are discussed and deliberated (Estafette, 2016).[8]

Furthermore, all four retailers emphasize the authenticity of food that is produced with no synthetic colouring, flavouring or preservatives, as well as its highly nutritious value. Thus, for Ekoplaza it is "food like food is intended" (Ekoplaza, 2016); for Marqt it is "real food"; for Natuurwinkel it is "healthy, beautiful, tasty and honest" food;[9] and for Estafette it is "high quality food for a healthy agriculture and fair economy".[10]

Social dimension

In the social dimension, all four retailers highlight the proximity with their suppliers in the form of direct personal relationships. In contrast, issues of race, class and culture (including traditional knowledge in farming practices) are not raised by the retailers examined here.

Ekoplaza notes that quality is decided daily together with suppliers, with whom it shares respect for each other and for the surrounding environment as well as a belief in the power of good food. Even in cases where producers are not in the local area, Ekoplaza underlines its effort to shorten the distance on the basis of transparency and by including only two steps in the supply chain: from the individual producer (e.g. in Peru, Sri Lanka or Moldavia) to the farmers' cooperative and then via Rotterdam port to its own packaging facility in Barneveld.[11]

Similarly, for Marqt the establishment of personal relationships with all farmers is fundamental (OneWorld, 2015). Marqt also communicates farmers' stories about their products on its website, thus aiming to create proximity with consumers on the basis of transparency.[12] Commenting on similar practices adopted by corporate retailers, Marqt notes that these create a false sense of familiarity and locality, e.g. by depicting Italian women happily cooking tomato sauce on a corporate brand label, while in fact the sauce is produced in large factory settings.[13] Furthermore, in order to create a closer relationship between producer and customer Marqt invites all supply chain actors during the opening of a new shop (Lommen, 2013).

Estafette, in turn, highlights its cooperative structure with the participation of various stakeholders, including final consumers. It also presents its producers on a separate webpage with links to its own website, though not in a storyline form. Natuurwinkel, in contrast, communicates more about its product assortment rather than the suppliers behind these products.

Environmental dimension

In the environmental dimension alternative food retailers emphasize different elements. The most encompassing is that of Ekoplaza which includes aspects of the sustainability of food production (as part of organic agriculture), animal welfare and consideration for future generations. In addition, Ekoplaza forms partnerships with other environmentally active actors, e.g. the Plastic Soup Foundation, an advocacy group that tackles plastic pollution, and Sungevity, a solar installation company. More specifically, in 2015 Ekoplaza signed an agreement with the Plastic Soup Foundation to support each other's mission. One of Ekoplaza's main ambitions is to become a plastic free supermarket. As part of this effort, Botanique, the cosmetic house brand, carries the label "Zero Plastics Inside" to communicate to consumers its commitment to the battle against microplastics in cosmetic and hygiene products.

Ekoplaza has also collaborated with two artists, McCardy and Hansky, in creating an exhibition, "The Plastic Supermarket", on the effects of plastic and displaying material collected from the Ij sea. Furthermore, Ekoplaza uses green energy in its retail shops and collaborates with Sungevity to give free advice to consumers on installing solar panels on their roofs. Likewise, through projects such as "geldGROENwassen", Ekoplaza customers can invest 250 euro in solar panels for animal farms and receive, in return, 300 euro in shop credit (Biojournaal, 2013).

Marqt emphasizes environmental sustainability at the farm level, as well as waste reduction and avoidance of food waste. In contrast to the other three retailers, however, who rely on certification, particularly regarding their food products, Marqt argues that certification is not the main means of demonstrating environmental sustainability. Indeed, it notes that although some farmers are not certified they personally guarantee the environmental attributes of their products. Thus, "environmental quality is not religiously defined through labels" (OneWorld, 2015).

For Natuurwinkel, the most important elements in the environmental dimension are the fostering of organic and biodynamic food production, seasonal food, locally produced (as far as possible) and with respect for humans, animals and nature. Furthermore, Natuurwinkel has adopted a "climate neutral" retail policy, paying particular attention to energy use, using "clean energy" for trucks and recycled paper for printing. It also provides tips for consumers on how to avoid food waste on its website.

Finally, Estafette prioritizes organic and biodynamic agriculture as well as seasonal products, while noting the formation of some (unspecified) relationship with the Slow Food movement. Estafette also organizes excursions to farms to familiarize consumers with organic farming.

Political dimension

In the political dimension, the question is whether a new form of agency for supply chain actors emerges as a result of a different organization of the supply chain. There is little information available to assess this dimension but some preliminary observations can be made.

First, all retailers emphasize the establishment of stable relationships with their suppliers based on trust and mutual respect. This, in turn, may create interdependence and mutual alliance in respect of future goals. Thus, Ekoplaza and its parent company Udea note that the closer contact between producers and wholesalers in organic food production generates interdependence and fair trade (Udea, 2013b). Due to its cooperative structure, Estafette Odin (2016) also maintains an interdependent relationship with its suppliers: "By founding Cooperation Odin, all people who want to work together in guiding food supply and food production in a healthy direction, are united: entrepreneurs, providers of capital, consumers, employees, producers and other stakeholders." However, more evidence is necessary in order to understand the extent to which supplier agency has been strengthened as a result of alternative food retailers.

With regard to consumers, alternative food retailers provide the space to reject the conventional food regime and embrace instead a different food system. In this context, studies demonstrate that while alternative food consumers have as their primary motivation personal taste and health concerns, they also share broader environmental and social values in respect of a fairer agri-food system that they want to express through their consumption behaviour (Seyfang, 2006). Likewise, such consumers tend to be motivated by behaviour that supports the development of "democratic, socially and economically just, and environmentally sustainable food systems" – the prime definition of food citizenship according to Wilkins (2005: 271). However, only Estafette Odin actively gives consumers the chance to influence retail strategies more directly by being members of the cooperative and having their voice heard through customer panels (Estafette Odin, 2016).

At the collective level, alternative food retailers (with the exception of Marqt) organize themselves in Bionext, the umbrella organization for sustainable food supply chains in the Netherlands bringing together the interests of farmers, traders, retailers and consumers. Bionext consists of three subdivisions: Biohuis, the organization for farmers and gardeners; BioNederland, the organization for traders, with 90 members representing 70 per cent of the organic market in 2015; and Biowinkelvereiniging representing 50 per cent of the 350 specialized retailers.[14] Biowinkelvereiniging members must have at least 90 per cent of their products certified as organic and pay an annual fee of 300 euro. Bionext is a member of the International Federation of Organic Agriculture Movements (IFOAM), representing the interests of the organic sector in the EU and at international fora.

Conclusion

This chapter has argued that justice is a core demand in the agri-food system. Although there is no common understanding of justice in the literature, the chapter has argued that it can be conceptualized as a multidimensional concept consisting of the following dimensions: material (access to safe and nutritious food); social (labour, gender, race and cultural concerns); environmental (disproportionate negative environmental impacts on the poor and also, more broadly, human ethical obligations to non-human species and future generations); and political (agency, the ability to substantively participate in, critically reflect upon and/or set one's own rules in the governance of food).

The chapter provided a critical evaluation of corporate models of intensive food production in regard to all the dimensions of justice and questioned whether alternative models, particularly in the form of alternative food retailers, can provide an adequate response. In this context, there are encouraging signs, but more evidence is required to provide a clearer answer. More specifically, in the material dimension, while access to safe and nutritious food is a core concern this is not expressed in the context of vulnerability to hunger. Rather, it is about making healthier and safer choices and providing a fair price to the farmer that is not determined by cost-efficiency but by a vision of a more sustainable agri-food system of "quality" and "authenticity". In the social dimension, the main focus is again the farmer and the ability of consumers to familiarize themselves with the person(s) behind the product. Issues of gender, race and class inequality are not raised by the retailers examined here. Likewise, traditional knowledge and its use in farming practices was also absent from the retail reports. In the environmental dimension, alternative food retailers adopt a rather comprehensive attitude to environmental sustainability, emphasizing not only sustainability at the farm but also at the retail shop and the supply chain as a whole. Nonetheless, with the exception of Marqt, the main evidence for environmental sustainability at the farm level is certification, which could result in the exclusion of non-standardized forms of sustainability and of farmers who cannot afford to demonstrate adherence to organic norms. In the political dimension, alternative food retailers provide space for the expression of food citizenship by giving consumers the opportunity to protest against the conventional food system. Yet, only in cooperative organizational structures are consumers given the opportunity to really participate in decision-making processes and influence the future of and vision for an alternative food system.

In sum, alternative food retailers recognize some, but not all, requirements for agri-food justice. Yet, despite limitations, they constitute an important element in transforming the conventional food system towards a fairer alternative.

Acknowledgements

I would like to acknowledge Bauke Ketelaar for valuable research assistance.

Notes

1 www.fao.org/food-loss-and-food-waste/en/.
2 The overarching system of public and private institutions that are valid in a given area of politics (Biermann *et al.*, 2009).
3 www.fao.org/3/a-bc270e.pdf.
4 www.oecd.org/agriculture/wateruseinagriculture.htm.
5 www.fao.org/save-food/resources/keyfindings/en/.
6 www.ekoplaza.nl/pagina/geen-corporatisme.
7 www.marqt.com/blog/2011/11/nederland-in-ideeen-2015/3459/.
8 www.estafette.org/pagina.php?sectie=missie.
9 www.natuurwinkel.nl/over-natuurwinkel.
10 www.estafette.org/pagina.php?sectie=missie.
11 www.ekoplaza.nl/pagina/eerlijke-handel.
12 www.marqt.com/blog/.
13 http://elsevierjuist.nl/bedrijf-belicht-marqt-wil-geen-elitaire-hang-out-zijn/.
14 https://bionext.nl/hoe-wij-werken.

References

Agarwal, B. (2014). Food Sovereignty, Food Security and Democratic Choice: Critical Contradictions, Difficult Conciliations. *Journal of Peasant Studies* 41(6), 1247–1268.

Alkon, A.H. and Agyeman, J. (2011). *Cultivating Food Justice. Race, Class and Sustainability*. Boston, MA: MIT Press.

Alonso, A. (2011). Farmers' Involvement in Value-Added Produce: The Case of Alabama Growers. *British Food Journal* 113(2), 187–204.

Altieri, M.A. (2009). Agroecology, Small Farms, and Food Sovereignty. *Monthly Review* 61(3), 102–113.

Angotti, T. (2012). The Urban–Rural Divide and Food Sovereignty in India. *Journal of Developing Societies* 28(4), 379–402.

Arias, P., Hallam, D., Krivonos, E. and Morrison, J. (2013). Smallholder integration in changing food markets. Food and Agriculture Organization of the United Nations, Rome.

Bacon, C.M. (2010). Who Decides What is Fair in Fair Trade? The Agri-Environmental Governance of Standards, Access and Price. *Journal of Peasant Studies* 37(1), 111–147.

Besky, S. (2014). *The Darjeeling Distinction. Labor and Justice on Fair-Trade Tea Plantations in India.* Berkeley, CA: University of California Press.

Biermann, F., Pattberg, P., van Asselt, H. and Zelli, F. (2009). The Fragmentation of Global Governance Architectures: A Framework for Analysis. *Global Environmental Politics* 9(4), 14–40.

Biojournaal (2013). Ekoplaza "wast geld groen" om Amsterdamse kinderboerderijen te voorzien van zonnepanelen. Available at: www.biojournaal.nl/artikel/13650/Ekoplaza-wastgeld-groen-om-Amsterdamse-kinderboerderijen-te-voorzien-van-zonnepanelen (accessed 15 June 2016).

Brown, S. and C. Getz. (2011). Farmworker Food Insecurity and the Production of Hunger in California, In: A.H. Alkon and J. Agyeman (eds.), *Cultivating Food Justice: Race, Class and Sustainability*. Boston, MA: MIT Press, pp. 121–146.

Castoriadis, C. (1991). *Philosophy, Politics, Autonomy*. Oxford: Oxford University Press.

Clapp, J. (2014). Financialization, Distance and Global Food Politics. *Journal of Peasant Studies* 41(5), 797–814.

Dauvergne, P. (2008). *The Shadows of Consumption*. Cambridge, MA: MIT Press.

Deda, P. and Rubian, R. (2004). Women and Biodiversity: The Long Journey from Users to Policymakers. *Natural Resources Forum* 28(3), 201–204.

Detailhandel (2016). *Biologische levensmiddelenzaken*. Available at: http://detailhandel.info/index.cfm/branches/levensmiddelenzaken/biologische-levensmiddelenzaken/ (accessed 8 May 2016).

Ekoplaza (2016). *Missie en Visie Ekoplaza*. Available at: www.Ekoplaza.nl/pagina/over-Ekoplaza-de-biologische-supermarkt-van-nede (accessed 9 June 2016).

Epstein, G. (2005). Introduction: Financialization and the World economy. In: G. Epstein (ed.), *Financialization and the World Economy*. Cheltenham, UK: Edward Elgar, pp. 3–16.

Estafette (2016). *Winkels*. Available at: www.estafettewinkel.nl/pagina.php?sectie=winkels (accessed 8 May 2016).

Estafette Odin (2016). *Over Estafette Odin*. Available at: http://estafette.org/pagina.php?sectie=over (accessed 15 June 2016).

European Parliament (2009). Food prices in spotlight amid concerns over supermarket domination. News release, reference no. 20090320STO52217. Available at: www.europarl.europa.eu/sides/getDoc.do?pubRef=-//EP//NONSGML+IM-PRESS+20090320STO52217+0+DOC+PDF+V0//EN&language=EN (accessed 14 September 2018).

FAO (2008). *The State of Food and Agriculture*. Rome: FAO.

FAO (2011). *Global Food Losses and Food Waste. Extent, Causes and Prevention*. Rome: FAO.

FAO (2012). The State of Food Insecurity in the World. Rome: FAO.

FAO (2014). The State of Food Insecurity in the World. Rome: FAO.

FAO (2016). *Migration, Agriculture and Rural Development: Addressing the Root Causes of Migration and Harnessing its Potential for Development*. Rome: FAO.

Foley, J.A., Ramankutty, N., Brauman, K.A., Cassidy, E.S., Gerber, J.S., Johnston, M., Mueller, N.D., O'Connell, C., Ray, D.K., West, P.C., Balzer, C., Bennett, E.M., Carpenter S.R., Hill, J., Monfreda, C., Polasky, S., Rockström, J., Sheehan, J., Siebert, S., Tilman, D. and Zaks, D.P.M. (2011). Solutions for a Cultivated Planet. *Nature* 478, 337–342.

Forman, F. and Mackie, G. (2013). Introduction: New Frontiers in Global Justice. *Critical Review of International Social and Political Philosophy* 16(2), 151–161.

Fraser, N. (2000). Rethinking Recognition. *New Left Review* 3, 107–120.

Fuchs, D. and Kalfagianni. A. (2014). Private Food Governance. In: C. Coff and P. Kemp (section editors on Governance and Policy), *Encyclopedia of Food and Agricultural Ethics*. Hamburg: Springer Academic Publishers, Article ID: 307765, Chapter ID: 259.

Fuchs, D., Glaab, K., Kalfagianni, A. and Meyer-Eppler, R. (2012). The Challenge of Food Security. In: R. Rayfuse and N. Weisfelt (eds.), *The Challenge of Food Security: International Policy and Regulatory Frameworks*. Cheltenham, UK and Northampton, MA: Edward Elgar Publishing, pp. 275–291.

Goodman, D., DuPuis, E.M. and Goodman, M.K. (2012). *Alternative Food Networks: Knowledge, Practice and Politics*. London and New York: Routledge.

Gottlieb, R. and Joshi, A. (2010). *Food Justice*. Cambridge, MA: MIT Press.

Gunders, D. (2012). Wasted: How America is Losing up to 40 percent of its food from farm to fork to landfill. National Resources Defense Council Issue Paper, August 2012, IP: 12-06-B.

Guthman, J. (2007). The Polanyian Way? Voluntary Food Labels as Neoliberal Governance. *Antipode* 39(3), 456–478.

Harward, E. (2010). Supermarkets' food pricing attacked. *The Dominion Post*, 4 July 2010. Available at: www.stuff.co.nz/national/3882634/Supermarkets-food-pricing-attacked (accessed 1 September 2018).

Hinrichs, C. and Kremer, K.S. (2002). Social Inclusion in a Midwest Local Food System Project. *Journal of Poverty* 6(1), 65–90.

Horrigan, L., Lawrence, R.S. and Walker, P. (2002). How Sustainable Agriculture Can Address the Environmental and Human Health Harms of Industrial Agriculture. *Environmental health perspectives* 110(5), 445–456.

Johnston, J., Biro, A. and MacKendrick, N. (2009). Lost in the Supermarket: The Corporate-Organic Foodscape and the Struggle for Food Democracy. *Antipode* 41(3), 509–532.

Kalfagianni, A. (2014). Addressing the Global Sustainability Challenge: The Potential and Pitfalls of Private Governance from the Perspective of Human Capabilities. *Journal of Business Ethics* 122(2), 307–320.

Kalfagianni, A. (2015). "Just Food": The Normative Obligations of Private Agrifood Governance. *Global Environmental Change* 31, 174–186.

Keuning, W. (2015). Hoge prijs remt groei biologische supers. Available at: http://fd.nl/ondernemen/1113626/hoge-prijs-remt-groei-biologische-supers (accessed 8 May 2016).

Le Mare, A. (2008). The Impact of Fair Trade on Social and Economic Development: A Review of the Literature. *Geography Compass* 2: 1–21.

Lockie, S. (2009). Responsibility and Agency within Alternative Food Networks: Assembling the "Citizen Consumer". *Agriculture and Human Values* 26, 193–201.

Lommen, G. (2013). Marqt consument & producent. Available at: www.Marqt.com/static//uploads/files/press/FoodPersonality_4202.pdf (accessed 1 September 2018).

Lyon, S. and Moberg, M. (eds.) (2010). *Fair Trade and Social Justice. Global Ethnographies*. New York: New York University Press.

McEntee, J.C. (2011). Realizing Rural Food Justice: Divergent Locals in the Northeastern United States. In: A.H. Alkon and J. Agyeman (eds.), *Cultivating Food Justice: Race, Class and Sustainability*. Boston, MA: MIT Press, pp. 239–259.

Marqt (2016). *Echt Eten*. Available at: www.Marqt.com/echt-eten/ (accessed 3 May 2016).

Natudis. (2016). *Over Natudis*. Available at: www.natudis.nl/over-natudis.php (accessed 15 June 2016).

Natuurwinkel. (2016). *Winkels*. Available at: www.natuurwinkel.nl/winkels (accessed 8 May 2016).
Nussbaum, M.C. (2000). Aristotle, Politics and Human Capabilities: A Response to Antony, Arneson, Charlesworth, and Mulgan. *Ethics* 111(1), 102–128.
NVM (Nederlandse Vereniging van Makelaars en Taxateurs in onroerende goederen) (2015). Vastgoedmarkt in Beeld.
O'Kane, G. (2016). A Moveable Feast: Exploring Barriers and Enablers to Food Citizenship. *Appetite* 105, 674–687.
OECD (2010). *Sustainable Management of Water Resources in Agriculture*. Paris: OECD.
OneWorld. (2015). Marqt: Niet alleen voor hipsters. Available at: www.oneworld.nl/food/eten-bedrijf/marqt-niet-alleen-voor-hipsters (accessed 25 September 2016).
Pearson, D., Henryks, J., Trott, A., Jones, P., Parker, G., Dumaresq, D. and Dyball, R. (2011). Local Food: Understanding Consumer Motivations in Innovative Retail Formats. *British Food Journal* 113(7), 886–899.
Rawls, J. (1971). *A Theory of Justice*. Boston, MA: Harvard University Press.
Runhaar, H.A.C., Melman, Th.C.P. Boonstra, F.G., Erisman, J.W., Horlings, L.G., de Snoo, G.R., Termeer C.J.A.M., Wassen, M.J., Westerink, J. and Arts, B.J.M. (2016). Promoting Nature Conservation by Dutch Farmers: A Governance Perspective. *International Journal of Agricultural Sustainability* 15(3), 264–281. Available at: http://dx.doi.org/10.1080/14735903.2016.1232015 (accessed 1 September 2018).
Sage, C. (2003). Social Embeddnedness and Relations of Regard: Alternative "Good Food" Networks in South West Ireland. *Journal of Rural Studies* 19, 47–60.
Santini, F., Gomez, S. and Paloma, G. (eds.) (2013). Short food supply chains and local food systems in the EU. A state of play of their socio-economic characteristics. EUR 25911 EN, Joint Research Centre, European Commission, Luxembourg.
Sen, A.K. (1999). *Development as Freedom*. New York: Knopf Press.
Seyfang, G. (2006). Ecological Citizenship and Sustainable Consumption: Examining Local Organic Food Networks. *Journal of Rural Studies* 22(4), 383–395.
Singh, S. (2003). Contract farming in India: Impacts on women and child workers. Gatekeeper Series No. 111, International Institute for Environment and Development, London.
Smit, P. (2012). Marqt vindt nieuwe investeerder. Available at www.levensmiddelenkrant.nl/nieuws/handel/formules/Marqt-vindt-nieuwe-investeerder (accessed 15 June 2016).
Steffen, W., Broadgate, W., Deutsch, L. and Ludwig, C. (2015). The Trajectory of the Anthropocene: The Great Acceleration. *The Anthropocene Review* 2(1), 81–98.
Stenmarck, Å., Jensen, C., Quested, T. and Moates, G. (2016). Estimates of European food waste levels. Fusions EU project, Swedish Environmental Research Institute, Stockholm.
Thanassoulis, J. (2008). Europe's farmers do not reap the benefits of higher food prices. *Politico*, 6 April 2008. Available at: www.politico.eu/article/europes-farmers-do-not-reap-the-benefits-of-higher-food-prices (accessed 1 September 2018).
Tilman, D. and Clark, M. (2014). Global Diets Link Environmental Sustainability and Human Health. *Nature* 515, 518–524.
Tsegaye, B. (1997). The Significance of Biodiversity for Sustaining Agricultural Production and Role of Women in the Traditional Sector: The Ethiopian Experience. *Agriculture, Ecosystems & Environment* 62(2–3), 215–227.
Udea (2013a). *Over Udea*. Available at: www.udea.nl/nl/over-udea/over-udea.html (accessed 15 June 2016).

Udea (2013b). *Mission & Vision.* Available at: www.udea.nl/en/about-udea/mission-vision.html (accessed 15 June 2016).
Wettstein, F. (2009). *Multinational Corporations and Global Justice.* Stanford, CA: Stanford University Press.
Wilkins, J.L. (2005). Eating Right Here: Moving from Consumer to Food Citizen. *Agriculture and Human Values* 22, 269–273.
Wittman, H.W. (2009). Reworking the Metabolic Rift: La Via Campesina, Agrarian Citizenship and Food Sovereignty. *Journal of Peasant Studies* 36(4), 805–826.

10 Social justice in the market place

The renewal of peri-urban open-air food markets around Montpellier, France

Coline Perrin, Elodie Valette and Claire Cerdan

Introduction

Farmers' markets are recognized as places of community building (Brown, 2002; Renting *et al.*, 2003). Numerous scholars have described direct sales as a means of developing solidarity between farmers and consumers and addressing the growing disconnect between production and consumption of food in urban systems (Morgan *et al.*, 2006). Localizing the food system is not an end in and of itself (Born and Purcell, 2006), but geographic and organizational proximity can meet the rising demand of city dwellers to know where and how their food is produced who has produced it. Promoting farmers' markets can thus be a way to sustain peri-urban agriculture and to reconnect producers and consumers. However, this producer/consumer relationship has often been idealized (Le Velly and Dubuisson-Quellier, 2008), leading scholars to question the actual impact of farmers' markets or other short food supply chains on social justice (Alkon, 2008; Mundler, 2013). "Farmers' markets may not provide farmers or their workers living wages, and farmers' markets may be less accessible, for a variety of economic and cultural reasons, to some communities" (Horst, 2017).

In this chapter, we analyze the renewal of open-air food markets in fast-growing peri-urban localities around Montpellier, a medium-sized city in the south of France. We discuss the impact of this renewal on the relationships between city dwellers and farmers and on social justice issues. Are open-air food markets equally accessible to both producers and consumers? Do these markets enable producers to obtain fairer payments and give all city dwellers access to quality food? Can local authorities contribute in specific ways to making markets fairer?

In order to contribute to this debate, this study focuses on the 31 municipalities which form the Montpellier metropolis, plus three adjacent municipalities. In this city-region, we identified 35 markets; 14 were in Montpellier (250,000 inhabitants) and 21 were in the peri-urban area (200,000 inhabitants). We carried out participant observation in the 21 peri-urban markets to gather information about the basic characteristics of each market (such as location, number of stalls, types of vendors, days of operation, range of goods offered), ran 169 quick surveys among markets food vendors (asking about the origin of the goods and the share of their sale in various supply chains) and conducted 27 in-depth interviews with vendors, mayors and technical staff in Montpellier and the peri-urban municipalities as well as with farmers' organizations. Through these interviews, we obtained a deeper understanding of the modus operandi of each market, and especially of the way local authorities perceive and manage markets. In particular, the in-depth interviews with vendors were focused on markets' governance and this highlighted issues of distributive and procedural justice between producers and resellers of food products.

This chapter is organized as follows. The first part consists of a literature review on food markets in France. The second part reveals our findings on the renewal of peri-urban open-air food markets around Montpellier and on their diversity. The third part discusses the impact of these open-air food markets on social justice for producers and consumers.

Recent interest in food markets in scientific literature

Though research on food markets has historically been found lacking (Zasada, 2009), scholars have recently shown a new interest in this topic, especially open-air food markets in the Global North in the context of the renewal of food markets, first in the US and then in Europe (Holloway and Kneafsey, 2000).

A primary focus of recent research concerns the role of markets as places of direct sale (Chaffotte and Chiffoleau, 2007; Schmutz-Poussineau, 2008). Scholars have analysed markets as alternative food networks, arguing that open-air markets, and especially producers' or farmers' markets, bring about new relationships between consumers and producers (Chiffoleau, 2008; Blumberg, 2018; Faraco, 2009; Navarro, 2011). They have also studied the change in consumers' perceptions and practices (Dury, 2008). Farmers' markets have indeed been considered "the historical flagship of local food systems" (Brown and Miller, 2008). The English-language literature focuses on the impact of farmers' markets on communities (Kirwan, 2004; Schmit and Gómez, 2011), showing that places of direct sale promote a sense of community, provide fresh food for consumers and yield positive economic impacts for local businesses (Gillespie *et al.*, 2007, Mont and Nilsson, 2017; Oberholtzer and Grow, 2003). French scholars, on the other hand, have focused on diversity among retailers: vendors, producers, artisans (Delfosse and Navarro, 2011). The issue of governance arises both in the French and English-language literature (Delfosse and Navarro, 2012; Derkzen, 2012; Dumain and Maurines, 2012; Machell and Caraher, 2012), along with the more general issue of rescaling food governance (Sonnino, 2016).

A second focus is related to the geography of food. Scholars have assessed the contribution of farmers' markets to an urban food strategy (Machell and Caraher, 2012) and highlighted the need to include markets within urban projects in relation to the planning of other commercial activities (Girard, 2011).

Geographical information systems (GIS) also help planners choose the best location for setting up new markets (Poisson, 2010; Scheffer and Dalido, 2010).

While French researchers have yet to deal with the issue of spatial access to food places, numerous works in the English-language literature have studied so-called food deserts, aiming to characterize areas where affordable and nutritious food is hard to access (Widener *et al.*, 2013). They have shown that food markets and retailers are unevenly distributed between high- and low-income areas. Beyond the spatial component, some scholars have advocated integrating a time component into research on food accessibility (Widener and Shannon, 2014) or considering the role of other informal food markets (Battersby, 2012). In his study of farmers' markets in the US, Schupp (2017) showed that there were social, economic and racial differences between those living in areas with farmers' markets and those in areas that did not have them.

Within this general context, French research first considered food markets in big cities and, more recently, has shown how mid-sized cities try to maintain their role as market places (Arnal, 2012; Marty, 2014). However, very few scholars so far have dealt with food markets located in peri-urban areas (Hugues-Dit-Ciles 2010). In Montpellier, where a recent study focused on the inner city's markets (Egret *et al.*, 2013), we decided to study peri-urban open-air food markets in order to reframe the issue of the contribution of these markets to the relocalization of food systems in the specific peri-urban context, where automobility has long shaped consumers' purchasing habits and facilitated the supremacy of supermarkets.

Open-air food markets around Montpellier

In this section, we explain the increasing number of open-air food markets around Montpellier and highlight their diversity.

The rise of open-air food markets around Montpellier

The number of open-air food markets has recently increased around Montpellier. In 2013 (see Figure 10.1), we identified two projects to create new markets and 33 existing markets, 12 of which are located in the city of Montpellier and 21 in peri-urban settings. Among these 21 peri-urban markets, seven have been created since 2000 (Table 10.1). This rise supports the recent trend in a renewal of markets in France (Navarro, 2015; Scheffer, 2010). This is especially significant in peri-urban areas where many markets, especially farmers' markets, disappeared in the 1970s with the massive increase in private car use and the opening of large-scale retail facilities.

This increase, however, is not uniform. We identified three enabling factors for the presence of a market in a peri-urban locality: distance to Montpellier, population size/dynamic and social profile of the locality.

First, the spatial distribution of the markets (Figure 10.1) shows that all the municipalities bordering Montpellier except one host a market, while most of

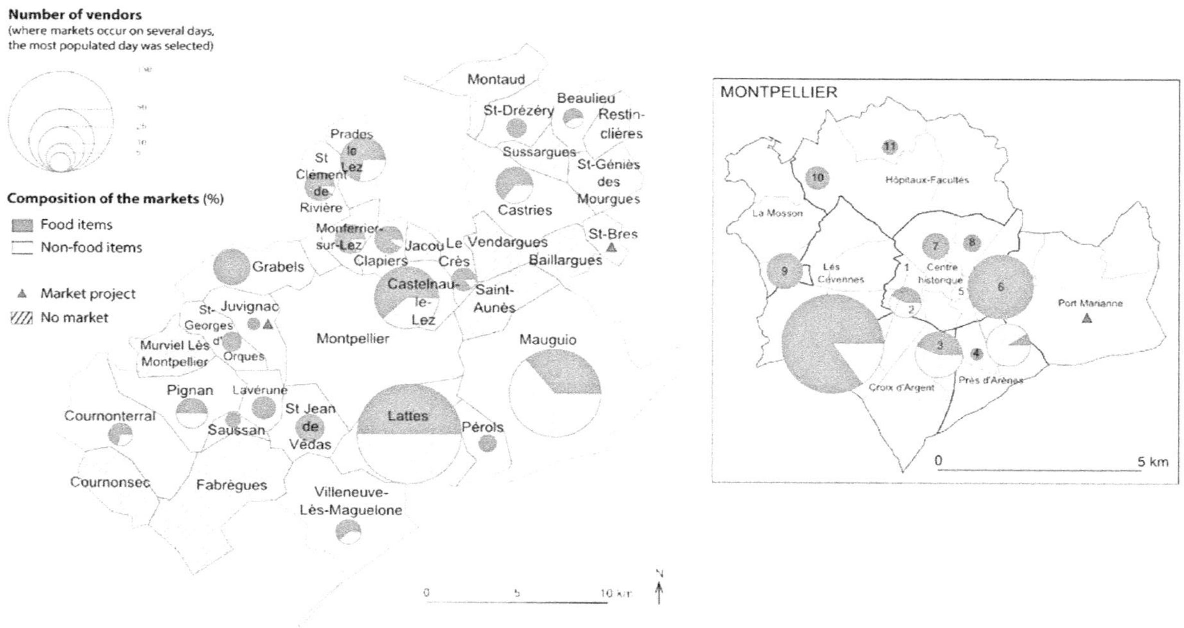

Figure 10.1 Location and size of markets in and around Montpellier (2013).

Note

Montpellier neighbourhoods and names of markets: (1) Les Arceaux – marché des Arceaux; (2) Figuerolles – marché Salengro; (3) Lemasson – marché Tastavin; (4) Saint Martin – marché Saint Martin; (5) Comédie – marché de la Comédie; (6) Antigone – marché d'Antigone; (7) Beaux-Arts – marché des Beaux-Arts; (8) Boutonnet – marché Albert Ier; (9) Celleneuve – marché de Celleneuve; (10) Hôpitaux Facultés – marché de Malbosc; (11) Plan des 4 Seigneurs – marché du Plan des 4 Seigneurs.

the municipalities without markets are located on the outskirts of the metropolitan perimeter. In the case of Prague, Spilková *et al.* (2013) show that the differences between markets "follow the inner city/hinterlands divide rather than the socio-spatial differentiation of the city". However, in the case of Montpellier, the spatial criteria may be correlated with the demographic and socio-economic criteria discussed below.

Population dynamics are a second key factor: all municipalities with more than 10,000 inhabitants host a market, compared with only 43 per cent of those with less than 2,500 inhabitants (see Table 10.1). The threshold for hosting a market is around 6,000–7,000 inhabitants. Furthermore, the most

Table 10.1 Characteristics of peri-urban municipalities and markets

Location	*Date of establishment*	*Number of stalls*	*Population*
Mauguio	1975–1980	100	16,307
Lattes	1980s	116	15,927
Castelnau-le-Lez	1950	50	14,948
Villeneuve-lès-Maguelone	1945	8	9,141
Saint-Jean-de-Védas	2001	10	8,716
Pérols	1970s	1	8,509
Le Crès	2000	14	8,005
Juvignac	1970s	2	7,466
Grabels	2008	17	6,393
Pignan	1993	12	6,366
Fabrègues	N/A	N/A	6,236
Baillargues	N/A	N/A	6,206
Cournonterral	1945	4	5,939
Castries	1966–1970	17	5,752
Vendargues	N/A	N/A	5,579
Saint-Georges-d'Orques	1992	5	5,343
Clapiers	2010	10	5,200
Saint-Clément-de-Rivière	1990s	11	4,987
Jacou	N/A	N/A	4,902
Prades-le-Lez	2009	20	4,534
Montferrier-sur-Lez	2000	11	3,385
Saint-Aunès	N/A	N/A	3,041
Lavérune	2006	7	2,735
Saint-Brès	N/A	N/A	2,659
Sussargues	N/A	N/A	2,567
Cournonsec	N/A	N/A	2,389
Saint-Drézéry	2008	5	2,156
Murviel-lès-Montpellier	N/A	N/A	1,911
Beaulieu	1980s	5	1,698
Saint-Geniès-des-Mourgues	N/A	N/A	1,692
Restinclières	N/A	N/A	1,554
Saussan	1996	3	1,482
Montaud	N/A	N/A	906

Note
N/A - data not available.

populated towns tend to have the biggest markets, with the highest number of vendors. Finally, though we did not find a clear correlation between the creation of a market and a town's demographic growth, this correlation is used as an argument by stakeholders: the establishment of new residential areas bring discussions about the opportunity to create a new market.

A third correlation can be suggested between the existence of markets and the presence of an older and wealthier population in peri-urban towns. The municipalities that host a market have a larger proportion of people over the age of 60 (more than 25 per cent) and retired people are over-represented in the markets' clientele (CCI de Montpellier, 2001). Nevertheless, vendors observe an increasing number of younger customers. The population's wealth, rather than age, appears to be a strong enabling factor

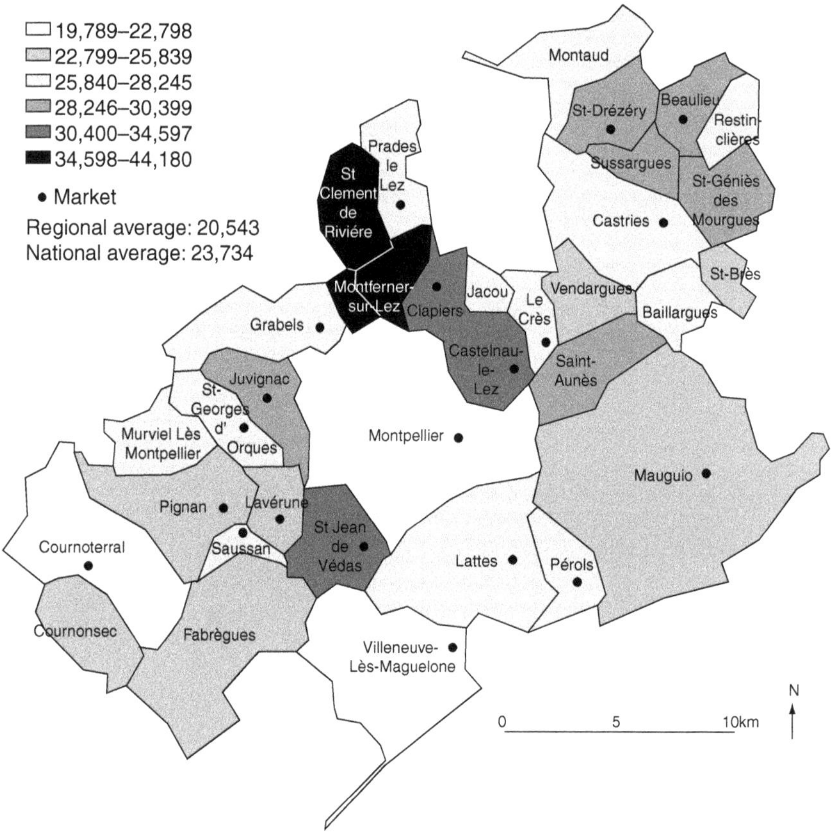

Figure 10.2 Average taxable income in and around Montpellier (2010).

Source: Directorate-General of Taxes, CCI Montpellier (used with permission).

Note

The income includes income from employment, income from capital and occasional (or irregular) income. It is obtained by dividing total household taxable income (in euros) by the number of taxable households.

(Figure 10.2). The recently created markets are located in municipalities with the highest taxable incomes in the north of Montpellier. On the other hand, the older markets are located in municipalities with average taxable incomes (south of Montpellier). This fiscal distribution is consistent with that of the socio-economic classification: a large proportion of middle- and high-income professionals live in the north of Montpellier, while low-income professionals and the working class live in the south and west of the city (Tailhades and Tasqué, 2010). The wealthier population supports and legitimizes the creation of new markets. Indeed, local authorities as well as some citizens' groups argue that this population's buying power and demand for quality products are responsible for the market's success.

In summary, new markets have emerged since 2000 in municipalities bordering the city of Montpellier (spatial criteria) with more than 6,000 inhabitants (demographic threshold) and a population older and richer than the metropolitan area's average. These social and economic enabling factors help us to understand the spatial distribution of markets, yet it is also important to highlight the diversity of the markets.

A large variety of open-air markets, on which producers are often a minority

Not all the markets are large, dynamic, or even held on a regular basis. They have a wide range of profiles, among which farmers' markets are a minority. In the Montpellier area, we identified four types of markets according to location, the date the market was established, the number of stalls and the presence of producers:

1. large, old peri-urban markets that offer food and non-food stalls, as well as an equivalent share of farmers and food retailers; these markets are located close to farming areas (Mauguio, Lattes, Castelnau);
2. large, old city markets, oriented towards organic products sold by farmers (Arceaux) or conventional products (Salengro), located within Montpellier, which draw in a wide variety of customers and have great appeal (customers come from remote areas);
3. small or very small local peri-urban food markets that were established some time ago (20–40 years), the continuation of which is not always ensured (Beaulieu, Juvignac);
4. new markets in wealthy and steadily growing peri-urban municipalities; these food markets are often farmers' markets and/or organic markets that organize special events and are strongly supported by residents (Montferrier-le-lez, Prades-le-Lez, Grabels).

Beyond this typology, our interviews with vendors tend to show the limited role of open-air food markets in maintaining peri-urban agriculture for two reasons: (i) vendors are not mainly agri-food producers; and (ii) markets are just one short food supply chain (SFSC) that farmers use in combination with others.

The number of producers on each market appears to be a good indicator of the profile of the market. Among the 169 vendors interviewed on peri-urban markets, one-third sold their own products, mainly fruit and vegetables, but also seafood (oysters), eggs, honey and wine. However, the number of producers varies (Figure 10.3): it was as high as 56 per cent of vendors in Grabels and 40 per cent in Prades-le-Lez, two wealthy municipalities where markets have been established in the past ten years with a clear objective by local authorities to attract farmers more than retailers and to favour SFSCs. These producers do not necessarily come from the surrounding areas. Proximity depends on the product: fruit and vegetables come from closer areas than meat or cheese.

Markets represent a variable share of sales for producers. It was the only point of sale for 3 producers out of 60 interviewed, and for another 3 producers it represented about 80 per cent of sales. However, most of the producers combine sales on markets with other SFSCs: direct sales on farms (wine makers), to restaurants (oysters) and to food retailers (fruit and vegetables). Some producers are also involved in community supported agriculture (CSA) schemes and producers' cooperative stores.

The rise of food markets in urban areas is often correlated in the literature with a demand for stronger links between producers and consumers. However, our results show that the rise of markets around Montpellier is rather correlated to spatial, demographic and socio-economic criteria and does not only concern farmers' markets. The extent and quality of the

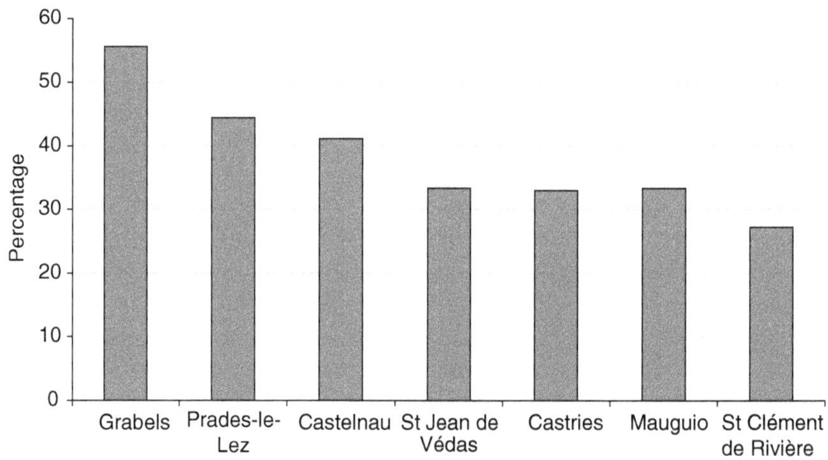

Figure 10.3 Share (per cent) of producers on peri-urban markets with more than ten vendors (2013).

relationships between producers and consumers at the market also depends on municipal policies.

Food markets in municipal policies: an issue of social justice?

In this section, we show the key role of municipalities in the management of the markets and discuss the impact of food markets on social justice, first, from the vendors' perspective and second, from the point of view of consumers.

Struggles in the market place: the supply side of justice

Food markets are places of economic competition. Struggles in the market place occur mainly between producers and retailers, and also among producers.

Perceived (in)justice between producers and retailers

As open-air food markets take place in public spaces in France, local authorities are the pivotal actor in their organization: they select the vendors and collect a fee – which varies greatly among municipalities and may depend on the size of the stall, the products sold (food or non-food) or even, sometimes, the status of the vendor (whether or not a producer). Local authorities also decide the spatial configuration of the markets, i.e. where stalls should be located and what facilities are required (access to water, electricity, etc.).

The distributive equity of municipal decisions is often criticized. Each municipality is free to dictate its own rules. On very small markets, every new vendor is welcome. On the contrary, on large successful markets there is competition among vendors keen to enter the market.

Most municipalities distinguish between regular vendors, who pay an annual fee (about 50 to 100 euro per linear metre) and have their stalls booked, and occasional vendors, who fill the remaining stalls. Stall allocation rules are specific to each municipality and not always written down. When there is a waiting list, the date of registration on the list is not the only criteria. Coming regularly as an occasional vendor is often a condition for obtaining a permanent stall. The type of products sold also counts. In Montpellier, for example, "the Mayor may depart from the order of the waiting list and give priority to a vendor whose activity is not or is under-represented on the market."[1] Some local authorities also impose strict quotas for stalls selling products that are already available in nearby permanent stores.

On some markets, priority is given to food vendors (100 per cent quota in Grabels, 60 per cent in Lattes), to agri-food producers (on one market they may even benefit from a fee exemption) or perhaps to local agri-food producers. In Grabels, for example, the market's charter requires products to be grown or processed within a 150 km radius (Bazin, 2013). In Mauguio, the municipality strongly supports local farmers: Mauguio's producers are clustered in two identifiable rows, whereas other producers are dispersed among the retailers of food and non-food products. The Mauguio locality still has numerous farmers, who often sell directly on farms or in huts on the side of the road. The market thus does not change the local supply of fruit and vegetables. Yet local farmers sometimes fear competition from the market, as well as from food stores located around the market place. This is why the municipality gives priority to local farmers willing to enter the market.

Unsurprisingly, the issue of how stalls are allocated leads to numerous conflicts and resentment among vendors, especially between producers and retailers. Beyond distributive equity, procedural justice is also criticized.

Participatory governance to improve procedural justice

Vendors often criticize the way municipalities manage food markets. They denounce the lack of information and transparency about the markets' rules. Important decisions are often taken by the mayor or the municipality's technical staff without public hearings or consultation with stakeholders. This may be considered a failure of procedural justice.

To address this issue, some municipalities set up a market committee so there can be dialogue with local stakeholders prior to the mayor's decision. Such committees are a first step towards participatory governance of the markets. However, vendors also criticize the market committees: their frequency (usually between one to six times a year) is not considered sufficient and they sometimes fail to represent all stakeholders. In the three recently created farmers' markets, the committees include elected officials and the municipality's technical staff and vendors (occasional vendors in three localities, local food stores in two localities), as well as consumers' representatives (three localities) and sometimes external experts (two localities). Some municipalities regret the lack of involvement of market vendors. This is partly due to the French national regulation that excludes individuals who are not "representative delegates belonging to a professional organization".[2] The Vendors' Union advocates for retailers, whereas producers are supported by the Association of Farmers' Markets and the Chamber of Agriculture. As market vendors do not always belong to such organizations, they may de facto be excluded from the committees.

Similar issues arise during the establishment of the market: a new market project often relies on one leader, often from the municipal staff. Only the three recently created farmers' markets and one other (Saussan) were set up by a working group that included various stakeholders. The priority given to producers was the key issue in the discussions. Some elected officials wanted a producers only market in order to address citizens' demands for local quality products. However, this preference does not always ensure the success of markets as it is difficult to ensure producers' regular presence. For instance, in

Prades-Le-Lez, one organic farmer started his business on public land. In return, the lease agreement specified that he was to sell his products at the market every Sunday. A few years later, having developed other profitable and less time-consuming ways of selling (CSA, sales to other producers), he stopped coming every week. Besides, because producers may not have a large range of products to sell during certain periods of the year, local authorities often welcome retailers as well, in spite of their initial preference for a farmers' market. In Clapiers, an elected member of the local authority dreamt of a farmers' market "like in the old markets, with a lot of colours and numerous producers" but admits he had "to be realistic [...]. Clapiers counts only 5,000 inhabitants. So, it is difficult to attract farmers to our starting-up market. For success, the market had to welcome producers and retailers" (interview conducted 17 July 2013).

The priority given to producers and the composition of the market may also be an issue of justice for consumers. This is discussed in the next section.

Improving the local foodscape: social justice for consumers

Food markets are promoted by municipal authorities as a means of diversifying the local food supply and promoting community building within comprehensive local food strategies.

Diversifying the local food supply: access to quality food for all?

Markets improve the peri-urban foodscapes when they complement or strengthen the local food supply. In Prades-Le-Lez, only one store continues to sell fruit and vegetables among the three that were running in the 1980s. The new food market offers fruit and vegetables, fish and ethnic products that are not sold in the local food stores. However, in Saussan, where the setting up of the market aimed to "[revitalize] the commercial activity of the municipality", the market now only counts three regular stalls (interview with an elected official from the municipality of Saussan, conducted 22 June 2013). Several food vendors left after a few years on account of a lack of sales in this small locality.

Besides the food supply itself, access to food also depends on physical accessibility and on prices.

Morning markets on weekdays exclude working people. To overcome this issue, some municipalities have an additional market in the evenings. Such schedules, however, may not fit the working practices of vendors. The difficulties in accessing the market by car (a lack of close parking spaces) may also exclude people living outside the village or those facing mobility issues (disabled or elderly people, pregnant women, etc.). To overcome these problems, one municipality has organized a bus service from the market to the car park and another has considered the idea of offering shopping trolleys (as in supermarkets) to help people carry heavy items.

Prices vary considerably from one market/vendor to another. Municipalities are reluctant to select only producers and vendors of high quality or organic products because they fear excluding lower-income residents and thus supporting the establishment of a "gourmet ghetto". Prices that appear fair to producers may be considered high by consumers and so exclude large segments of the population. This issue of social justice for consumers explains why diversified food markets, which welcome both producers and retailers, are easier to start and to develop than farmers' markets in small localities. Similar justice issues have been raised in relation to farmers' markets in the US (Alkon, 2008). They have also been discussed in respect of other SFSCs (Mundler, 2013).

Finally, diversity on the market will be fair for producers and retailers as well as for consumers if, and only if, the origin and methods of production and processing are clearly displayed. This is often not the case. Most vendors resell food products that they have bought at wholesale markets or directly from farms. Customers, therefore, do not know where, and by whom, each product has been produced and processed. Delfosse and Navarro (2012) point to this recurring lack of transparency, even on the part of producers who do not always sell only their own products. In an attempt to rectify this, an action-research project involving the French National Institute for Agricultural Research (INRA) and the Grabels municipality developed labels with different colours giving information about who has produced and processed the products (Chiffoleau *et al.*, 2013). Informing consumers about the typicality of products and providing accountability with regard to their origin are levers to improve trust and social justice on food markets.

Figure 10.4 Coloured labels for food products on Grabels' market.

Source: V. Sancelme, 2013.

Community building at the market place

Beyond food procurement, the market may improve the local foodscape through community building.

Kirwan (2004) refers to social connectivity, reciprocity and trust as the dominant motivations of British consumers who buy from farmers' markets. In future research, further interviews with consumers would be necessary to discover whether most customers are only interested in the act of purchasing – thus, not spending very much time at the market and considering it an ordinary shopping place – or if the market is rather a place where community building really happens.

Local authorities do believe markets contribute to community building. According to one municipal employee, the social function of the market depends on its economic success: "The more stalls there are, the more costumers and social bonds there will be" (interview with an economic affairs officer from Montpellier, conducted 23 July 2013). Indeed, large, old markets are lively places. They are an entertainment in themselves and do not need side activities to attract customers.

At smaller markets, local authorities or grassroots organizations promote markets as places for social cohesion, especially between the generations.

Figure 10.5 The market – a place of conviviality.

Source: V. Sancelme, 2013.

They organize tastings or other events (raffles and lottery games, culinary contests and cooking classes, fanfare or music, carriage rides, etc.). The mayor of Clapiers wanted to create "a friendly place" where "families may come to stroll around and not only to purchase". He suggested that time spent at markets that take place at weekends is considered leisure time (interview conducted 17 July 2013). In Prades-Le-Lez, the mayor aimed at "building a place that attracts and retains customers, a meeting place, a place for social and intergenerational diversity" (interview conducted 9 July 2013). Mayors' interest in markets is also due to other reasons: the competition between localities; and a friendly market is a good way to promote the village and attract tourists. Markets are also the best places for mayors to meet their constituents!

Finally, markets may also be the starting point for local projects linking agriculture and the city, improving the knowledge that city dwellers have of agriculture and farming. By making local agri-food more visible in public spaces, markets "educate customers on the potential for and seasonal limits of local food" (Brown and Miller, 2008). The Association of Farmers' Markets, for instance, organizes visits for young people "because it is important to see the producers, the market and the cook [...] to strengthen friendly ties and develop a taste for good products" (interview with the President of the Association of Farmers' Markets in Languedoc-Roussillon, conducted 11 June 2013). In Montpellier, schools visit markets once a year where they

Figure 10.6 Inauguration of the evening farmers' market in Grabels.

Source: V. Sancelme, 2013.

cook with a chef. In Grabels, a teacher decided to integrate the market in the curriculum: children asked the producers for some seeds, grew their own vegetables, and sold them at the market. At the end of the school year, the annual school dance also took place at the market. Many other projects have emerged with consumers, and the municipality has supported farmers willing to start a business, created urban gardens, changed food procurement channels for school canteens, etc. Finally, in the three municipalities that have set up a farmers' market in the last ten years, the market has become a starting point for a territorial strategy on agriculture and food.

Nevertheless, while they were generally concerned about the social inclusivity of their markets, the municipalities we interviewed did not have explicit objectives in respect of disadvantaged populations. We could not identify any financial incentives or tools designed to enhance the purchasing power of lower-income households, whether at the national or the local level. There are many initiatives of this kind in the US where, for example, vouchers are provided to beneficiaries for use at farmers' markets (Dimitri *et al.*, 2015) to help increase access to and the affordability of healthy agri-food (Valentine *et al.*, 2017). These incentives have proved to be effective in improving nutrition among recipients of US federal food assistance, especially when combined with exposure interventions (tasting, cooking, etc.) at farmers' markets (Bowling *et al.*, 2016).

Conclusion

To conclude, the renewal of open-air food markets around Montpellier appears to be socially selective, as is the case for many short food chain initiatives. Since 2000, new markets have emerged, particularly in municipalities bordering the city with more than 6,000 inhabitants and populations older and richer than the metropolitan area's average. The results presented in this chapter, however, are only a snapshot of a selection of peri-urban farmers' markets. They would need to be complemented by longitudinal studies in order to assess the evolution of the percentage of producers on markets and the exact relation of markets with surrounding agriculture.

The impact of markets on social justice was mainly discussed in relation to the way markets are managed by municipalities. From the vendors' perspective, as well as that of consumers, the most important justice issue relates to the presence of both producers and retailers at French food markets. However, struggles and perceived injustices arise between them concerning the allocation of available stalls. Beyond distributive equity, procedural justice was also discussed: vendors' resentments are exacerbated by the lack of transparency and participation in the municipal management of the markets. The support of local authorities and the participation of consumers, retailers and producers in market governance, through the creation of a market committee, for example, are two essential levers to transform French open-air food markets into places of community building and improved social justice.

Besides these governance issues, the impacts of markets on sustainable food practices and on social justice for consumers and producers are also directly linked to the way local authorities articulate the market as an element of their local food strategy, and should be investigated as such.

Acknowledgements

The authors would like to thank Virginie Sancelme for her assistance in conducting the survey and analysing the survey results.

Notes

1 Article 10, Réglement Général des Marchés de Montpellier [municipal regulation].
2 Article 3, Réglement type des Marchés de France (2011) – Fédération Nationale des Syndicats des Commerçants des Marchés de France.

References

Alkon, A.H. (2008). From Value to Values: Sustainable Consumption at Farmers Markets. *Agriculture and Human Values*, 25(4), pp. 487–498.

Arnal, C. (2012). La place et le rôle de l'agriculture à la périphérie des villes moyennes. Le cas des villes d'Annecy, Bourg-en-Bresse, Montbrison et Romans-sur-Isère. PhD Thesis, Aménagement et Urbanisme, Université Lumière, Lyon.

Battersby, J. (2012). Beyond the Food Desert: Finding Ways to Speak about Urban Food Security in South Africa. *Geografiska Annaler: Series B, Human Geography*, 94(2), pp. 141–159.

Bazin, M. (2013). Quelles médiations marchandes et sociales pour favoriser le développement des circuits courts alimentaires? Diagnostic et perspectives à partir de l'expérimentation d'un nouveau système d'étiquetage sur un marché de plein vent. Diplôme d'Ingénieur. Master's dissertation, Agrocampus Ouest, Rennes.

Bléchet, C., Cinquin, M. and Vant, A. (1985). Eléments pour une Géo-Histoire des Marchés Forains Lyonnais. *Revue de géographie de Lyon*, 60(1), pp. 109–128.

Blumberg, R. (2018). Alternative Food Networks and Farmer Livelihoods: A Spatializing Livelihoods Perspective. *Geoforum*, 88, pp. 161–173.

Born, B. and Purcell, M. (2006). Avoiding the Local Trap: Scale and Food Systems in Planning Research. *Journal of Planning Education and Research*, 26, pp. 195–207.

Bowling, A.B., Moretti, M., Ringelheim, K., Tran, A. and Davison, K. (2016). Healthy Foods, Healthy Families: Combining Incentives and Exposure Interventions at Urban Farmers' Markets to Improve Nutrition among Recipients of US Federal Food Assistance. *Health Promotion Perspectives*, 6(1), p. 10.

Brown, C. and Miller, S. (2008). The Impacts of Local Markets: A Review of Research on Farmers Markets and Community Supported Agriculture (CSA). *American Journal of Agricultural Economics*, 90(5), pp. 1298–1302.

Brown, A. (2002). Farmers' Market Research 1940–2000: An Inventory and Review. *American Journal of Alternative Agriculture*, 17, pp. 167–176.

CCI de Montpellier (2001). Direction "Etudes et Observatoire Economique". Les halles et marchés – livre blanc. Available at: www.montpellier.cci.fr/pages/index.php?id_m=&id_arbo=007604570464&men=&id_page=338#halles.

Chaffotte, L. and Chiffoleau, Y. (2007). *Les marchés classiques de plein, ouverts toute l'année*. Fiche Commercialisation CROC No. 2. Montpellier. INRA. Available at: www.google.com/url?sa=t&rct=j&q=&esrc=s&source=web&cd=1&ved=2ahUKEwjL16eA0qbdAhXozIUKHQ8TASYQFjAAegQIAxAB&url=https%3A%2F%2Fwww.plate-forme21.fr%2Fspip.php%3Faction%3Dacceder_document%26arg%3D1013%26cle%3D5e5509b2a8a23d8b949e3bd82807448586dae521%26file%3Dpdf%252Fcouverture_cahiers.pdf&usg=AOvVaw24-6vAJ00g_xHedohW2zH7.

Chiffoleau, Y. (2008). Les Circuits Courts de Commercialisation en Agriculture: Diversité et Enjeux pour le Développement Durable. In: G. Maréchal (ed.), *Les Circuits Courts Alimentaires: Bien Manger dans les Territoires*. Dijon: Educagri, pp. 21–30.

Chiffoleau, Y., Dalmais, M. and Divet, J-P. (2013). Signaler et garantir l'origine des produits dans les circuits courts alimentaires: L'expérience du marché de Grabels. Série Les Focus PSDR3, Projet PSDR, Languedoc-Roussillon. Available at: www.psdr.fr/archives/INS33PDFN8.pdf.

Delfosse, C. and Navarro, A. (2011). Spécificités et Renouveau des Marchés dans le Cadre des Circuits Courts. *Les Carnets Pro de Liproco*, 9, p. 12.

Delfosse, C. and Navarro, A. (2012). Nouveaux Marchés, Nouvelles Gouvernances. In: A.H. Prigent-Simonin and C. Hérault-Fournier (eds.), *Au Plus près de L'assiette: Pérenniser les Circuits Courts Alimentaires*. Versailles: Quae; Dijon: Educagri, pp. 187–198.

Derkzen, P. (2012). Food and the City : The Challenge of Urban Food Governance. In: J. Wiskerke and A. Viljoen (eds.), *Sustainable Food Planning: Evolving Theory and Practice*. Wageningen, The Netherlands: Wageningen Academic Publishers, pp. 61–65.

Dimitri, C., Oberholtzer, L., Zive, M. and Sandolo, C. (2015). Enhancing Food Security of Low-Income Consumers: An Investigation of Financial Incentives for Use at Farmers Markets. *Food Policy*, 52, pp. 64–70.

Dumain A. and Maurines B. (2012). Composer les Manières de Gouverner. In: A. Prigent-Simonin and C. Hérault-Fournier C. (eds.), *Au Plus près de L'assiette, Pérenniser les Circuits Courts Alimentaires*. Dijon: Educagri, pp. 216–231.

Dury, S. (2008). Les consommateurs de l'Hérault face aux circuits courts de commercialisation et aux produits locaux. *Cahiers de l'Observatoire CROC*, 5, Project CROC, Montpellier, p. 14.

Egret, L., Meneroud, P., Robert, C. and Souaiaia, H. (2013). Marchés à dominance alimentaire et leurs influences sur la ville de Montpellier. Master's dissertation, Université Paul Valéry de Montpellier.

Faraco, C. (2009). Les marchés de Lyon au travers des relations producteurs-consommateurs. Un enjeu dans l'approvisionnement de la ville en produits locaux. Master's dissertation, Université Lumière Lyon 2.

Gillespie, G., Hilchey, D.L., Hinrichs, C. and Feenstra, G. (2007). Farmers' Markets as Keystones in Rebuilding Local and Regional Food Systems. In: C.C Hinrichs and T.A. Lyson (eds.), *Remaking the North American Food System: Strategies for Sustainability*. Lincoln, NE and London: University of Nebraska Press, pp. 65–83.

Girard, P. (2011). *Structurer les marchés de plein vent. Pays Haut Languedoc et Vignobles*. Available at: www.payshlv.com/files/phlv/documents/PDF/Economie/ETUDE%20MDPV%202010%2018-01.pdf.

Holloway, L. and Kneafsey, M. (2000). Reading the Space of *the Farmers' Market: A Preliminary Investigation from the UK. Sociologia Ruralis*, 40(3), pp. 285–299.

Horst, M. (2017). Food Justice and Municipal Government in the USA, *Planning Theory & Practice*, 18(1), pp. 51–70.

Hugues-Dit-Ciles, J. (2010). Marchés de détail et producteurs dans le Pays Mornantais: Les enjeux du renouvellement des marchés dans un territoire périurbain. Master's dissertation, Université Lumière Lyon 2.

Kirwan, J. (2004). Alternative Strategies in the UK Agro-Food System: Interrogating the Alterity of Farmers' Markets. *Sociologia Ruralis*, 44(4), pp. 395–415.

Le Velly, R. and Dubuisson-Quellier, S. (2008). Les Circuits Courts entre Alternative et Hybridation. In: G. Maréchal (ed.), *Les Circuits Courts Alimentaires: Bien Manger dans les Territoires*. Dijon: Educagri, pp. 105–112.

Machell, G. and Caraher, M. (2012). The Role of Municipal Markets in Urban Food Strategies: A Case Study. In: J. Wiskerke and A. Viljoen (eds.), *Sustainable Food Planning: Evolving Theory and Practice*. Wageningen, The Netherlands: Wageningen Academic Publishers, pp. 127–136.

Marty, P. (2014). De la Restauration Scolaire à L'intégration de L'agriculture dans le Projet de Territoire: Vers une Ville Acteur Normatif de la Question Agricole? *Géocarrefour Revue de Géographie de Lyon*, 89(1), pp. 135–143.

Mont, O. and Nilsson, H. (2017). Socioeconomic Aspects of Farmers' Markets in Sweden. *System Innovation for Sustainability*, 3, pp. 113–128.

Morgan, K.J., Marsden, T.K. and Murdoch, J. (2006). *Worlds of Food: Place, Power and Provenance in the Food Chain*. Oxford: Oxford University Press.

Mundler P. (2013). Le Prix des Paniers Est-il un Frein à L'ouverture Sociale des AMAP? Une Analyse des Prix dans Sept AMAP de la Région Rhône-Alpes. *Économie rurale*, 336(4), pp. 3–19.

Navarro, A. (2011). La valorisation des produits locaux sur les marchés de détail du PSADER de l'agglomération lyonnaise: Limites et enjeux de l'offre locale. Master's dissertation, Université Lumière Lyon 2.

Navarro, A. (2015). Le marché de plein vent alimentaire et la fabrique des lieux: Un commerce de proximité multifonctionnel au coeur de la recomposition des territoires. PhD thesis, Université Lumière Lyon 2.

Oberholtzer, L. and Grow, S. (2003). Producer-only farmers' markets in the mid-Atlantic region: A survey of market managers. Henry A. Wallace Center for Agricultural & Environmental Policy at Winrock International, Arlington, VA. Available at: www.researchgate.net/publication/237111558/download.

Poisson, M. (2010). Observer les circuits courts à l'échelle d'un territoire proposition d'analyses spatiales des données en termes de système d'informations géographiques. Séminaire PSDR Grand Ouest, Rennes, 6 January 2010.

Renting, H., Marsden, T.K. and Banks, J. (2003). Understanding Alternative Food Networks: Exploring the Role of Short Food Supply Chains in Rural Development. *Environment and Planning A*, 35(3), pp. 393–412.

Scheffer, S. (2010). La Dimension Touristique des Marchés, une Forme Particulière de Renouveau. In: J-B. Traversac (ed.), *Circuits Courts: Contribution au Développement Régional*. Paris: Educagri, pp. 129–151.

Scheffer, S. and Dalido, A-L. (2010). Observer les circuits courts à l'échelle d'un territoire: Proposition d'un modèle d'analyse spatiale des données, en termes de systèmes d'information géographique. One hundred and sixteenth seminar, European Association of Agricultural Economists, Parma, Italy, 27–30 October 2010.

Schmit, T.M. and Gómez, M.I. (2011). Developing Viable Farmers Markets in Rural Communities: An Investigation of Vendor Performance Using Objective and Subjective Valuations. *Food Policy*, 36(2), pp. 119–127.

Schmutz-Poussineau, A-M. (2008). La Vente Directe sur les Marchés. In: G. Maréchal (ed.), *Les Circuits Courts Alimentaires: Bien Manger dans les Territoires*. Dijon: Educagri, pp. 31–42.

Schupp, J.L. (2017). Cultivating Better Food Access? The Role of Farmers' Markets in the US Local Food Movement. *Rural Sociology*, 82(2), pp. 318–348.

Sonnino, R. (2016). The New Geography of Food Security: Exploring the Potential of Urban Food Strategies. *The Geographical Journal*, 182(2), pp. 190–200.

Spilková, J., Fendrychová, L. and Syrovátková, M. (2013). Farmers' Markets in Prague: A New Challenge within the Urban Shoppingscape. *Agriculture and Human Values*, 30(2), pp. 179–191.

Tailhades, B. and Tasque, O. (2010). Les chiffres clés de Montpellier Agglomération. *Repères, Synthèse* (4).

Valentine, H., Harvey, S., Landfried, L., Craven, T., Brock, G. and Gibson, C. (2017). Customer Response to a Supplemental Nutrition Assistance Program (SNAP) Nutrition Incentive Program at Farmers Markets. *Journal of the Academy of Nutrition and Dietetics*, 117(9), p. A84.

Widener, M.J. and Shannon, J. (2014). When are Food Deserts? Integrating Time into Research on Food Accessibility. *Health & Place*, 30, pp. 1–3.

Widener, M.J., Metcalf, S.S. and Bar-Yam, Y. (2013). Agent-Based Modeling of Policies to Improve Urban Food Access for Low-Income Populations. *Applied Geography*, 40, pp. 1–10.

Zasada, K. (2009). Markets 21: A policy & research review of UK retail and wholesale markets in the 21st century. The Retail Markets Alliance. Available at: www.nabma.com/wp-content/uploads/2015/05/4dd6419ec9584_markets-21-report.pdf.

11 Protection of a place?

Community supported agriculture (CSA) in Germany

Birgit Peuker

Introduction

A distinctive characteristic of community supported agriculture (CSA) is that it attempts to establish cooperation between farmers and consumers. This is reflected in a specific business organization in which a group of consumers pre-finances the running operating costs of a farm in exchange for a part of its produce. The farmer is thereby insured against the risks of cultivation and price changes on agrarian markets. The objective of building this cooperation is justified by a critique of the modern agri-food-system. In this way, a political, value-based dimension is added to an economic concept of business organization. The CSA project as a farm is an economic entity as well as a political unit. This political character is not only expressed in the specific business organization but also in the contentious character of CSA projects: they not only want to be an alternative, they also want to speak it out loud – through the use of flyers and websites and by participating in political protest.

In Germany, CSA is not very widespread in comparison with other countries. According to the German CSA network around 129 projects or initiatives exist in Germany at present[1] – compared to the USA with around 1,144 projects and France with around 1,200 (in 2010) (Schlicht *et al.*, 2012: 22, 29). The interest in CSA projects in Germany among scientists, students and political actors is nevertheless high. One sign of this is that the German network of CSA farms established a working group to coordinate scientific activities, in order to align research questions and avoid overloading some CSA-farms.[2]

In Germany, many studies on CSA neglect their social movement character and conceive of CSA as a tool of innovative regional development (Kraiß and van Elsen, 2011). The political dimension of CSA is often disregarded. However, this dimension must be taken into consideration to explain why people are motivated to engage in CSA projects and why they establish CSA in specific ways. In this chapter, I discuss CSA against the background of social movement research. From this point of view CSA can be seen as part of a long tradition of critical voices on industrialized agriculture (see, for example, Rachel Carson, 1976) which was an element of a wider critique on industrialization more generally. Initially, the political character of CSA appears in the intention to be an "alternative" to the mainstream; it emphasizes the specific constructive dimension of CSA. The chapter initially addresses the issue of how CSA practices are constituted as an alternative to industrialized agriculture and support the political claims made by CSA projects. The second part of the chapter shows that social movement research emphasizes first, the role of "identity" as central to social movements and second, "self-administered companies" as the main collective political strategy. The third part discusses the core concept of CSA which gives diverse CSA farms an identity. The fourth part explores how organizing practices emerge around this core concept to do both: keep the project going and meet the requirements of the core concept.

CSA as a social movement

Many activists as well as social scientists conceive of CSA not only as an alternative way of marketing but as a political grassroots movement opposed to industrialized agriculture (see O'Leary, 2010; Charles, 2011; Press and Arnould, 2011; Schnell, 2013; Bîrhală and Möllers, 2014). Social movements are defined as a specific form of contentious politics. They emerged with early industrialization as a historical form to make collective claims in respect of a targeted authority by bundling together specific strategies, such as the founding of associations and protest marches. The movement presents itself to the public as strong, worthy of support and united, although internal differences might exist (see Tilly and Wood, 2009). Throughout their history the character of social movements has, however, changed. In the 1970s a new wave of social movements emerged, labelled by scholars as "new": e.g. the environmental movement, the peace movement and the feminist movement. The main critique of the environmental movement, in particular, was of industrialization and its impacts. The industrialization of agriculture also formed part of this critique (Carson, 1976). After World War II, agriculture modernized, in terms of the machinery used as well as the use of fertilizer and pesticides, in both West and East Germany. Carson (1976) – a US biologist and writer who was also read extensively in Germany – pointed to the environmental consequences of industrialized agriculture, especially the use of pesticides.

With the new social movements the way of doing politics changed; the new form was sometimes called "sub-politics" (Beck *et al.*, 1999). Contrary to the "old" politics, the aim was to change society not, initially, by making political claims (through political parties and parliament) but by changing one's own everyday practices as part of an alternative lifestyle. Consuming eco-farming products was part of this. With regard to "sub-politics", a key factor was not only individual practices but, above all, collective practices. In Germany, the "alternative movement" established many self-administered projects in the sphere of child care (*Kinderläden*), education, leisure and production as well as in agriculture (rural communes) (Heider, 2008; Reichardt, 2014: 324ff.). The aim was to establish community-based projects in opposition to the processes of individualization.

Despite the downturn of alternative economies in the 1990s (only a few projects survived), CSA can be seen as a revival of alternative projects. Interest in agriculture and in eco-farming was resurrected in the wake of anti-globalization movements. An increased awareness of rural problems by urban activists was the result of solidarity with Latin American peasant movements such as the Landless Workers' Movement and La Via Campesina which opposed neoliberalization. Simultaneously, German consumers increasingly came to distrust industrialized food production. The growing discomfort with modern food production was triggered by food scandals such as mad cow disease (bovine spongiform encephalopathy or BSE), the dioxin scandal and the rotten meat scandal. The 1970s and 1980s concept of an "alternative economy" was reformulated as "solidary economy" (Heider, 2008: 526). In the 2000s, peasant-based agriculture became a symbol for eco-social-transformation processes – not only in the view of German activists and scientists but worldwide (for Germany, Ötsch, 2016; Müller, 2002: 4; beyond Germany, see Duncan, 2000). Since 2010, annual protest marches have taken place on the occasion of the agricultural Green Week (*Grüne Woche*) in Berlin. With the slogan "*wir haben es satt*" ("we are fed up") diverse individuals, groups and organizations – such as Slow Food, FoodFirst Information and Action Network (FIAN) and Friends of the Earth – protest against industrialized food production and support animal rights and eco-farming. Around the same time, the number of CSA projects started to "upswing" (Bietau *et al.*, 2013: 8). Early studies on CSA in Germany speak of ten CSA farms by 2009 (Kraiß and van Elsen, 2011).

The main problem for social movement theory is explaining how collective action is possible and through which mechanisms individuals act collectively (Eder, 2015). With regard to CSA the question can be reframed as: Why do people engage in CSA? One of the basic concepts of social movement theory that is useful in answering this question is "identity". Identity helps to institutionalize a project or movement to unite different worldviews in a single protest or project. However, in respect of social movements more generally, Rucht (2011) pointed out that an identity must not be too strong, otherwise it will not enable activists from different social spheres to cooperate. Likewise, for CSA the importance of a guiding principle – which can be considered to be the identity – is emphasized (Asendorf *et al.*, 2003: 24).

Core concepts of CSA

The core concept of CSA is the idea of building trust-based cooperation between consumers and farmers. This idea is backed up by a special form of business organization. Pre-financing the running costs of a farm is interpreted as an expression of the confidence that consumers have in farmers. Consumers gain the possibility of participating in business decisions: for example, they may be involved in the decision about what crops to grow and how they will be grown. It is intended that within CSA old roles have to be dismantled. Farmers do not operate the farm on their own anymore; consumers help out on the farm and help to distribute the harvest. In this way, consumers and farmers become participants, shareholders (O'Leary, 2010) or co-farmers (*MitBäuerInnen*) (Künnemann and Presse, n.y.).

This core concept gradually became apparent in a slow grassroots process (Kraiß and van Elsen, 2011: 356). The history of CSA is often told like a myth in which the key person is Trauger Groh who lived on "the first" community farm in Germany, "*Buschberghof*". There, he learned about the idea that consumers and farmers could collectively organize a farm. Then, in the mid-1980s, he founded the first CSA farm in the USA. The idea quickly spread, and more and more farms were founded based on the idea of CSA. It was reimported back into Germany in the 2000s and the initial farm ("*Buschberghof*") still exists in its original form. The large variety of CSA farms is one expression of their grassroots character.

The importance of the name of the concept highlights the difficulties in translating the English term "community supported agriculture" into a German easy-to-handle label. At the founding meeting of the CSA network in Germany, the different world views and interests of those present became obvious. With the chosen term "*Solidarische Landwirtschaft*" ("solidary agriculture") it was possible to link different understandings. The support of new farms was accepted as a basic objective of the network (Peuker, 2011). But some projects still reject the term "*Solidarische Landwirtschaft*"; to emphasize the community-based dimension of their projects they speak, for example, of "*gemeinschaftstragende Landwirtschaft*" ("community-based agriculture") (Bauer, 2014) or "*Freihof*" (literally, "freefarm") (Künnemann and Presse, n.y.).

The core concept was not, however, only the result of social actors' efforts to shape it; it was also shaped by the social context.

First, CSA as a social movement is part of a broader movement which is opposed to industrialized agriculture. The main points of critique of industrialized agriculture in a broader sense are:

- *ecological risks*: for example, the use of pesticides, over-fertilization, water shortage, CO_2 pollution;
- *inequalities in the agri-food-system*: for example, hunger on the one side and obesity on the other; malnutrition, especially among poor people;
- *reduction of (agro-)biodiversity*: this is an important cross-over problem which combines ecological and food production risks – the loss of biodiversity threatens food production in the future;
- *reduction in the quality of food*: unhealthy and food lacking in taste, loss of contact with food production.

The last point of the broader critique of industrialized agriculture is especially stressed by CSA: the loss of contact with food production by consumers, the lack of a relationship between consumers and farmers and long transport routes in the industrialized agri-food system. The alternative proposed by CSA is a close relationship between consumers and food production embedded in a local community. Eco-farming is not a top-priority, although – unlike in other countries where non-eco-farming CSA projects exist (Soil Association, 2011; Möllers and Bîrhală, 2014) – CSA is deeply connected with eco-farming in Germany (Bietau *et al.*, 2013: 65).

Second, the method of constructing an alternative was inspired by social movement practices in the 1970s and 1980s labelled as "alternative economy" and, later on, "solidary economy". Part of this alternative economy is the self-administration of projects and companies. This politicization of business within a social movement has a tradition in Germany as well as in other industrialized countries in the West. With the "alternative movement" ("*Alternativbewegung*") in the 1980s, many self-administered companies emerged as part of an alternative economy. On the one hand, self-administered companies have a stabilizing effect on social movements, because of their long-term dimension compared to short- term protest events. On the other hand, the long-term nature of self-administered companies may result in a stronger adaption to the economic environment and a loss of the political dimension (Heider, 2008: 514). The same debate can be found in relation to CSA. Some activists claim that CSA without politics – that is, to establish the marketing concept without a trust-based community – is no longer CSA (Hayden and Buck, 2011: 333; Soil Association, 2011: 6, 12; McFadden, 2013: 94).

As studies on self-administered companies in Germany show, the motives for engaging in such projects or for supporting them by buying their products are political. The political dimension manifests itself in the intention to provide an alternative to contemporary society and to engage, not for individualistic cost–benefit reasons, but for post-materialistic values such as self-realization and community-building (Reichardt, 2014: 319).

Third, the fact that community building is thought to be political can only be understood if we look at new social movements in general. The main critique of new social movements is stated as being against individualization in modern society. Rationality, technology and social indifference is criticized; the emphasis is on subjectivity, emotions and concrete experience in a modest and ecological lifestyle (Reichardt, 2014: 189ff.). The main objective of new social movements is to develop a process of self-expression and acknowledgement on the basis of trust. Communities are considered "free associations" which anyone can leave if they disagree. As in the self-administered companies of the 1980s, community building and work are combined (Reichardt, 2014: 319ff.).

Not every CSA activist might be aware of this tradition, but the milieu of CSA activists borrows much from history. Besides post-materialistic values, being part of a community is a major reason for the intrinsic motivation to engage with CSA projects. Recent studies, in which CSA activists in different countries were asked about the reasons for their engagement, have shown that support for local businesses is an important reason to participate (Soil Association, 2011; Schnell, 2013: 621).

Fourth, being part of a network of farms is an important feature of a given CSA project. It allows farms to open up to the public, to promote the idea of CSA and to make a difference in local politics as well. There are local and nationwide networks of CSA projects – of new partnerships between consumers and farmers (Wild, 2012). A prominent example of a global CSA network is "URGENCI" which involves projects of consumer–producer partnerships that might be very different in many aspects, though they agree on common goals.[3]

Such networks help to promote the idea of CSA and therefore can be seen as part of the political dimension of CSA projects: they want not only to establish a farm but also to motivate others to do the same.

Collective action through CSA practices

In the previous section the core concept of CSA (*Solidarische Landwirtschaft*) was described as the intention of constructing a community which is supported by a specific operating model in which participants annually pre-finance the running costs of a farm. This core concept gives identity to the project members. With their concrete practices within CSA projects, they reshape and consolidate this identity. With respect to these current organizing practices, the following questions can be raised: How do CSA projects constitute themselves as an alternative to industrial agriculture? Through which organizational techniques and practices do CSA projects build the aspired community?

Since practices occur in time and place they differ within different CSA projects. The projects are diverse and not every project is the same, though there is some kind of best practice in CSA. It is possible to look at already generalized practices which can be found in manuals on CSA.[4]

Laying the cornerstones: farm and stakeholders

What does it mean to establish a project and to give birth to an enterprise in the long run? The first challenge is to find a farm, a farmer and a group of consumers (stakeholders). Usually, the farm should be close to where consumers live. It might be difficult to find a suitable farm, due to the availability of land on the agricultural land market (Künnemann and Presse, n.y.). In Germany, the agricultural land market is becoming increasingly monopolized, the number of small farms is decreasing and big farms are growing. In these circumstances, CSA is one way for small-scale farmers to continue in agriculture. If the farm has to be set up from scratch, initial investments are required. Agricultural machinery and equipment need to be bought. Another possibility is to apply for local government and state grants and subsidies (Hartkemeyer *et al.*, 2013). Likewise, decisions regarding what kind of products the farm will produce need to be taken. Which kind of crops are to be grown? Is livestock farming going to be part of the project? How big will the farm be? (see also Künnemann and Presse, n.y.). About two-thirds of farms in Germany produce vegetables, one-half produce fruit and only 12 per cent include meat as part of their production (Bietau *et al.*, 2013: 63*)*. This is similar to other Western countries such as the UK where vegetables are predominantly grown (Soil Association, 2011: 24). German CSA farms also tend to produce processed products, e.g. juices, spreads, bread and cheese (Bietau *et al.*, 2013: 63*)*.

At the same time, the bigger the farm and the further away it is from where stakeholders live the more difficult it is to build a trust-based community centred around a place. Although the community is connected to a specific place, this connection is not as strong as in a typical self-administered company or commune where members live and work together.

Legal obligation

Consumers and farmers often sign a private law contract, sometimes for one year or perhaps for 20 weeks (depending on the cultivation period). The contract specifies the contribution consumers have to pay, the cultivation techniques to be used and how the harvest is to be divided and distributed (Wild, 2012: 36; Künnemann and Presse, n.y.: 71). Some projects reject this legal framework as a way of conducting business relations and prefer the informal method of a shared economy. For example, it is possible to agree that financial contribution depends on what a stakeholder can offer or is willing to give (Künnemann, 2012: 71). It is important to fix the price of the weekly contribution because farm expenditure needs to be calculated correctly and income from contributions should be high enough to meet these financial commitments. Empirical studies of CSA farms in the USA have shown that only half of the farms surveyed were highly satisfied with how the operating costs were covered (see Brown and Miller, 2008: 1299). On German CSA farms, contributions are usually determined in agreement with stakeholders (Bietau *et al.*, 2013: 64).

These practices show that the community is temporally limited. There are no indications of any CSA projects that extend beyond a one-year timeline. One indicator of attachment to the CSA project is the number of annual contract renewals. The possibility of leaving the community after one year might be to the benefit of participants (and to the character of the CSA project as a "free association"), but holds a risk for the farm: it may not be able to find enough participants every year to maintain the project and the farm (Hartkemeyer *et al.*, 2013; for experiences outside Germany, see O'Leary, 2010: 18).

Farm representation group

A farm representation group is typically elected by all members of the project at an annual meeting. This supports the farmers by undertaking business decisions, management and public relations. However, what is prescribed in the manual rarely happens in practice (there are, perhaps, two or three manuals on CSA in German available on the internet to advise new farm groups setting up CSA farms). Whereas in other countries such as the USA there are only a few CSA projects with no farm representation group, in Germany the lack of a farm representation group is more common – especially on farms which are family based. Studies on German CSA farms show that only two-thirds of farms take collective decisions on what to grow, and the number reduces to around 40 per cent with regard to investments. In about 80 per cent of German CSA farms the individual farmer has the final say (Bietau *et al.*, 2013: 65).

In principle, in building a trust-based community a farm representation group is not needed in order to bridge the gap between farmers and consumers. It is possible to have trust-based relationships within a hierarchy or between different social positions. Beyond trust, however, there is a second possible feature of community: democracy and participation. Community is not conceived of in a traditional way, such as in regard to family, but as an "association of the free". The self-administered company levels hierarchies and empowers participants in respect of self-realization. In this context, CSA can be embedded in different world views: those towards the right on the political spectrum, which give a central position to the family that runs the farm, and those towards the left which emphasize reclaiming space and food sovereignty. In this sense, an alternative can mean being conservative and family based or progressive and commune based. On CSA farms in Germany one can find both forms of community orientation.

Volunteering

Taking part in farming practices is important in order to build a connection with plants and crops and other forms of non-human life on the farm. In this way, the community itself is enhanced. The "place" is filled with life through knowing about the provenance of food and how it is produced (Bauer, 2014: 198; the same is true for CSA projects outside Germany – see, for example, Schnell, 2013: 621; Bougherara *et al.*, 2009: 1494).

In Germany, volunteering on CSA farms is not common. Volunteering is usually an additional way to make contact with the farm and is widely used by members (Bietau *et al.*, 2013: 71).Volunteering requires agricultural skills; learning by doing often requires an expert, the farmer or other trained people, to mentor the volunteers. For example, volunteers have to learn to distinguish between plants and weeds so they do not remove the plants (Bauer, 2014: 198).

The learning that goes on conveys the idea that CSA is an alternative not only in its organizational form but also in the everyday practices of its members. Participants experience that by getting involved in volunteering they really are part of an alternative.

Sharing the harvest

For some activists, self-sufficiency is the key idea behind CSA projects. Even if not all the food needed by a household can be produced by a single CSA farm, the aspiration is to distribute the entire harvest between the participants (Künnemann and Presse, n.y.: 3). This is not always feasible, however. Some CSA projects buy-in other products so they have a wider variety of crops to distribute between participants. Other projects sell parts of the harvest that are not needed on farmers' markets or in farm shops (Bietau *et al.*, 2013: 66). Fruit and vegetables are only available on a seasonal basis and not all year round. In order to anticipate consumer demand, some manuals on CSA advise buying-in products which are not grown on the farm to ensure variety in weekly boxes (Wild, 2012).

The unwritten rule of self-sufficiency can be interpreted as a relic of the concept of commune within social movements, where living and working together was a standard feature. CSA seems to intend to renew this idea on a networked scale, which comprises rural and urban actors.

In German CSA projects participants are content with the quality of the food produced and only dissatisfied about the quantity of food to be shared (Bietau *et al.*, 2013: 77) On the one hand, the supermarket somehow remains a model for consumer demand; it shapes the expectations of participants. Moreover, participants need to readapt to the logic of seasonal agricultural production and to new types and varieties of food. In this respect, it is obvious that the dichotomy between consumers and producers is not dissolved by CSA projects: first, because different roles persist in society; and second, participants' expectations are shaped by the modern agri-food-system which not only disadvantages but also benefits the modern consumer. On the other hand, consumers embed CSA practices in their everyday shopping habits. Most CSA members buy the rest of the food they need as organic quality (Bietau *et al.*, 2013: 67).

The problems around sharing the harvest demonstrate the tensions created by different individual needs and experiences. These tensions arise because CSA projects are not a full-time community like a commune – they are not even a part-time community, but a sequential time community, where participants spend only a short part of their lifetime.

Distribution of the produce

In some CSA projects participants collect their share of the harvest from the farm on their own on a fixed day of the week. Other projects set up "pick-up-points" or organize car-sharing to transport produce (Bauer, 2014). In Germany, collection by the customer and pick-up-points are the most common ways of distributing the produce (Bietau *et al.*, 2013). The purpose of distributing boxes is not only to carry the agricultural produce to urban consumers but to help consolidate the trust-based community. If participants collect their share at the farm, they can meet and speak with each other or the farmer. Besides that, they can visit the farm and observe the farming practices. Even in the case of "pick-up-points" participants can meet one another and self-organize distribution.

But the goal of strengthening the community sometimes conflicts with ecological objectives. For example, some members may be concerned that if household members individually drive their own cars to the farm this may increase greenhouse gas emissions. Another argument against individual collections of shares is that people who cannot afford to drive a car or who reject car driving are excluded from the project (Künnemann, 2012).

Thus, two values which may be labelled as alternative guiding principles – community building and ecological concerns – may conflict. This suggests that that there is not a clear dichotomy between "mainstream" and "alternative": in a given society there may be many alternatives. CSA is concerned with community in the first instance, not with ecological thinking. However, the "alternative" in attitudes and practices becomes an identity in itself and facilitates the cooperation of people with different social and political backgrounds.

Organizing farm festivals and setting up newsletters

Manuals on CSA suggest that in order to establish or enhance a community something more has to be done than just farming or management. For example, farm festivals and public farm visits help the community to strengthen its cohesion. A basic measure is to share information through a website or by distributing newsletters, books and brochures (Wild, 2012: 49f.; Künnemann and Presse, n.y.). For example, the German CSA farm "Pente" portrays farm life in an e-mail newsletter "News from the farm", sent to its members at regular intervals (and later published in a booklet for the wider public). The diverse contents of these reports include weather information, aspects of cultivation, farmer colleagues and information about food politics and recent developments in agri-food production (Hartkemeyer *et al.*, 2013).

Beyond discussions around specific practices, the CSA concept is subject to intense discussion around the question whether CSA is open to everybody. CSA appears to attract specific types of people. Various studies report that CSA projects attract mainly women, white-collar workers, urban dwellers, people aged 30–40, those with children and those educated in higher education (for Pennsylvania, USA, see Schnell, 2013: 620; for China, see Chen, 2013: 39; for the UK, see Soil Association, 2011: 24). Some studies find that people of colour are under-represented in CSA projects, so that some speak of CSA as a "white space" (Charles, 2011: 364). These exclusion mechanisms are not intended but reflected in CSA projects. To address these concerns, handbooks on CSA provide examples of strategies to involve under-represented groups. Those on low incomes can be supported, for instance, by reducing their financial share and distributing the outstanding amount between the other participants ("share-a-share") (Künnemann and Presse, n.y.; 13, Soil Association, n.y.). Another proposal is to proactively get in touch with under-represented groups to find out the reasons why they do not participate (Soil Association, n.y.).

This discussion cannot be understood without the insights provided by social movement theory. CSA projects favour the value "open" over "closed", but why? New social movements aim at grassroots democratic organizing principles in order to level the power asymmetries in society. Cultural plurality and diversity of members are values in themselves. But as this discussion shows, social inequalities make it difficult to achieve this goal.

Conclusion

This chapter has examined the social phenomenon of CSA through the lens of social movement theory to answer the question whether and how CSA constitutes an alternative to the contemporary agri-food system. The chapter demonstrated that the intention to constitute an alternative is an authentic objective of new social movements and it is manifested in the political character of CSA projects. Building a community is essential to new social movements, in opposition to individualization processes; meanwhile, more conservative and more progressive notions of communities exist at the same time. The specific business model of CSA borrows much from the self-administered companies of former movements such as the alternative movement of the 1970s and 1980s.

For some activists CSA is an alternative way to abolish capitalism and with it the contemporary agri-food system. But for other CSA participants capitalism is not the problem; rather, they want the option to choose the kind of food they consume.

To paraphrase Eckert (2015): CSA is a practice movement in which practices themselves build a room, and reclaim and free participants. In this sense, CSA is part of alternative lifestyle concepts. It is unlikely that CSA will effectively nourish a large number of people. As Kraiß and van Elsen (2011) point out, experts estimate that CSA could feed around 10 per cent of the population in Germany. It is also not very likely that a large number of people will be highly motivated to take part in CSA projects. More importantly, perhaps, structural limitations demotivate extrinsically: lack of time, lack of access to land and inconsistencies with regard to ecological concerns. For these reasons, we expect that CSA will stay an alternative within a niche.

Notes

1 www.solidarische-landwirtschaft.org/de/solawis-finden/solawi-hoefe-initiativen/.
2 www.solidarische-landwirtschaft.org/de/mitmachen/arbeitsgruppen/forschung/.
3 http://urgenci.net/vison-and-mission/.
4 For examples of handbooks on how to develop CSA projects, see Wild, 2012; Künnemann and Presse, n.y.

References

Asendorf, I., Demmeler, M., Flieger, B., Jaudas, J., Sauer, D. and Scholz, S. (2003). Nachhaltigkeit durch regionale Vernetzung – Erzeuger-Verbraucher-Gemeinschaften im Bedürfnisfeld Ernährung. Endbericht des BMBF-Forschungsprojektes "Nachhaltigkeit durch regionale Vernetzung – Erzeuger-Verbraucher-Gemeinschaften im Bedürfnisfeld Ernährung" [Sustainability through regional networking – Producer–consumer communities in the field of need related to nourishment]. Report of a research project funded by BMBF: (German Ministry of Education and Research) and produced by Tagwerk A.G., Technische Universität München and Institut für Sozialwissenschaftliche Forschung – ISF-München.

Bauer, C. U. (2014). Solidarische Landwirtschaft – Modell für den Hof der Zukunft? Im Gespräch: Elmar Schulte-Tigges, Dipl.-Geograph und Betreiber des "Lernbauernhof Schulte-Tigges" und der "Erzeuger-Verbraucher-Gemeinschaft Kümper Heide" [Community supported agriculture – model of a farm of tomorrow? Interview with Elmar Schulte-Tigges, geographer and operator of the farm for teaching "Schulte-Tigges" and the producer–consumer community "Kümper Heide"]. *Standort* 38, 198–202.

Beck, U., Hajer, M. and Kesselring, S. (1999). Der unscharfe Ort der Politik – eine Einleitung [Blurry Place of Politics – An Introduction]. In: U. Beck, M.A. Hajer and S. Kesselring (eds.), *Der unscharfe Ort der Politik. Empirische Fallstudien zur Theorie der reflexiven Modernisierung*. Opladen, Germany: Leske and Budrich, pp. 7–22.

Bietau, P., Boddenberg, M., Dietze, F., Frauenlob, M.H., Gunkel, L., Kärgel, K., Leierseder, S., Munz, J., Schmitz, S., Sergan, N. and Vaessen, F. (2013). Solidarische Landwirtschaft – eine soziale Innovation? Eine empirische Studie aus soziologischer Perspektive [Community supported agriculture – A social innovation? An empirical sociological study]. Johann Wolfgang Goethe, Universität Frankfurt.

Bîrhală, B. and Möllers, J. (2014). Community supported agriculture in Romania. Is it driven by economy or solidarity? Discussion Paper No. 144, Leibniz Institute of Agricultural Development in Transition Economies.

Bougherara, D., Grolleau, G. and Mzoughi, N. (2009). Buy Local, Pollute Less: What Drives Households to Join a Community Supported Farm? *Ecological Economics* 68, 1488–1495.

Brown, C. and Miller, S. (2008). The Impacts of Local Markets: A Review of Research on Farmers Markets and Community Supported Agriculture (CSA). *American Journal of Agricultural Economics* 80(5), 1296–1302.

Carson, R. (1976). *Der stumme Frühling* [*Silent Spring*, first published 1962]. München: Beck.

Charles, L. (2011). Animating Community Supported Agriculture in North East England: For a "caring practice". *Journal of Rural Studies* 27, 362–371.

Chen, W. (2013) Perceived Value of a Community Supported Agriculture (CSA) Working Share. The Construct and Its Dimensions. *Appetite* 62, 37–49.

Duncan, C.A.M. (2000). The Centrality of Agriculture: History, Ecology and Feasible Socialism. *Socialist Register* 36, 187–205.

Eckert, J. (2015). Practice Movements: The Politics of Non-Sovereign Power. In: D. Della Porta and M. Diani (eds.), *The Oxford Handbook of Social Movements*. Oxford: Oxford University Press, pp. 568–575.

Eder, K. (2015). Social Movements in Social Theory. In: D. Della Porta and M. Diani (eds.), *The Oxford Handbook of Social Movements*. Oxford: Oxford University Press, pp. 32–45.

Hartkemeyer, J., Hartkemeyer, M., Hartkemeyer, J. and Hartkemeyer, T. (2013). *Nachrichten vom Hof. Lichtblicke zwischen Landlust und Landfrust* [*News from the Farm. Rays of Hope with Pleasure and Pain of Rural Life*]. Norderstedt, Germany: Books on Demand.

Hayden, J. and Buck, D. (2011). Doing Community Supported Agriculture: Tactile Space, Affect and Effects of Membership. *Geoforum* 43, 332–341.

Heider, F. (2008). Selbstverwaltete Betriebe in Deutschland [Self-Administered Companies in Germany]. In: R. Roth and D. Rucht (eds.), *Die Sozialen Bewegungen in Deutschland seit 1945. Ein Handbuch. Kapitel 22.* Frankfurt and New York: Campus Verlag, pp. 513–527.

Kraiß, K. and van Elsen, T. (2011). Community Supported Agriculture (CSA). Ein nachhaltiges Konzept für ländliche Räume [Community Supported Agriculture (CSA). A Sustainable Concept for Rural Regions]. Beitr. 11 Wiss.- Tagung Ökol. Landbau (Band 2). Gießen, pp. 356–359.

Künnemann, R. (2012). Die Markushofgemeinschaft in Maisbach [the title refers to a specific CSA project named "Markusgemeinschaft" in the village of Maisbach]. In: S. Wild (ed.), *Sich die Ernte teilen… Einführung in die Solidarische Landwirtschaft.* Heimsheim, Germany: Printsystem Medienverlag, pp. 70–71.

Künnemann, R. and Presse, M. (n.y.). Wir gründen einen Solidarhof. Leitfaden zur Solidarischen Landwirtschaft [Founding a CSA project. Manual for community supported agriculture].

McFadden, S. (2013). Der Kontext der Gemeinschaftshöfe [The Context of Community Farms]. In: G. Trauger and S. McFadden, with B. von Wolfgang Stränz and M.R Ostrom. *Höfe der Zukunft. Gemeinschaftsgetragene/Solidarische Landwirtschaft (CSA).* Darmstadt: Verlag Lebendige Erde, pp. 69–99.

Möllers, J. and Bîrhală, B. (2014). Community Supported Agriculture: A Promising Pathway for Small Family Farms in Eastern Europe? A Case Study from Romania. *Landbauforschung: Applied Agricultural and Forestry Research* 64(3/4), 139–150.

Müller, C. (2002). Zukunftsfähige Landwirtschaft eine Aufgabe nicht nur für Bauern [Not Only for the Farmer is Sustainable Agriculture a Task]. In: M. Wohlan (ed.), *Zukunft der Wirtschaft: Landwirtschaft und Ernährung.* Bonn: Bundeszentrale für politische Bildung, pp. 36–42.

O'Leary, D.E. (2010). Shifting landscapes: Community supported agriculture's next season. PhD thesis. Syracuse University.

Ötsch, S. (2016). Der Unternehmer ist tot – es lebe das Gärtnernde? Leitbilder und Wirtschaftskonzepte der Postwachstumsgesellschaft [The Entrepreneur is Dead – Long Live the Gardener? Guiding Principles and Economic Concepts in Post-Growth Societies]. *Momentum Quarterly* 5(1), 18–35.

Peuker, B. (2011). Solidarische Landwirtschaft [Community Supported Agriculture]. *Gen-ethischer Informationsdienst* 207, 19–21.

Press, M. and Arnould, E.J. (2011). Legitimating Community Supported Agriculture through American Pastoralist Ideology. *Journal of Consumer Culture* 11(2), 168–194.

Reichardt, S. (2014). *Authentizität und Gemeinschaft. Linksalternatives Leben in den siebziger und frühen achtziger Jahren* [*Authenticity and Community. Left-Leaning Alternative Lifestyles in the 1970s and early 1980s*]. Berlin: Suhrkamp.

Rucht, D. (2011). The Strength of Weak Identities. *Forschungsjournal Soziale Bewegungen*, 24(4), 73–84.

Schlicht, S., Volz, P., Weckenbrock, P. and Le Gallic, T. (2012). Community supported agriculture: An overview of characteristics, diffusion and political interaction in France, Germany, Belgium and Switzerland. ACT*eon* Environment Research & Consultancy and die Agronauten, Research for Sustainable Regional Economies.

Schnell, S.M. (2013). Food Miles, Local Eating, and Community Supported Agriculture: Putting Local Food in Its Place. *Agriculture and Human Values* 30, 615–628.

Soil Association (2011). The impact of community supported agriculture. Soil Association CSA Support Project, Making Local Food Work.

Soil Association (n.y.). Frequently asked questions on CSA.

Tilly, C. and Wood, L.J. (2009). *Social Movements, 1768–2008*. Boulder, CO and London: Paradigm Publishers.

Wild, S, (2012). Praxisteil [Practical Part]. In: S. Wild (ed.), *Sich die Ernte teilen, … Einführung in die Solidarische Landwirtschaft* [Sharing the Harvest. Introduction to Community Supported Agriculture]. Heimsheim, Germany: Printsystem Medienverlag, pp. 14–55.

12 Short food supply chain responses to agri-food system challenges

Concluding remarks

Agni Kalfagianni and Sophia Skordili

This book explores the potential of short food supply chains (SFSCs) to address the multiple challenges – environmental, social and economic – facing the global agri-food system today. Ranging from farmers' markets, to community supported agriculture (CSA), collective food procurement, food hubs, micro-entrepreneurship and alternative food retailers, SFSCs now provide (re)newed innovative forms of collective action in an effort to foster sustainable transformations at the local level within a globalized and increasingly unsustainable agri-food system. In this final chapter we return to the two broad analytical lenses – place and governance – in order to reflect on the promises and the pitfalls of SFSCs in attempting to foster a more sustainable agri-food system.

Place

In Chapter 2, Morgan and Santo observe that the "local" label can often conceal as much as it reveals. The principle of local is found, explicitly or implicitly, in any initiative to develop sustainable SFSCs. However, it is embued with varying degrees intensity, even different content, among the many different settings and types of actors in SFSCs.

Loose ties with the locality

The preceding chapters have shown that several agri-food products produced under SFSCs are addressed at the national or even international market. In several instances this is an indication of dynamism, e.g. Parmigiano-Reggiano; in others, the main motive is survival due to the low capacity of the local market, e.g. Feta-Thessalia. The broad geographical range on the consumption side is common practice in SFSCs. What is less obvious, though, is the concurrent extensive range of the production side.

Several chapters in this book show that SFSCs make use of supplementary sources of agricultural inputs. Some of them are within the boundaries of the local food system area, while others lay outside. This is not a surprise for bigger actors in supply chains. Kalfagianni (Chapter 9) shows that alternative retailers operating in the Netherlands make use of domestic suppliers as well as suppliers from far away destinations such as Peru, Sri Lanka and Moldovia, although they make an effort to reduce drastically the number of steps within the supply chain. However, this is less anticipated in respect of small local firms. Fieldwork research by Perrin *et al.* (Chapter 10) shows the limited supply from local farmers at the open-air markets operating in Montpellier peri-urban areas. This is attributed to the use of supplementary market channels by the vast majority of producers and, especially, to the over-presentation of traders rather than producers at the open-air markets. The survey of micro local agri-food firms located in the hinterland of the greater Athens area (Chapter 6) also shows that they make use of agricultural input from distant origins all over the country. In this case the defining factor is the search for lower price supplies.

Hence, local embeddedness, on both the consumption and production sides, of the main actors in SFSCs is rather loose. Morgan and Santo (Chapter 2) do not see this open character of the SFSC as a problem. In fact, the opposite could lead to "green parochialism".

The ties of SFSCs with the intangible assets of localities are correlated, to a large extent, with the historical trajectories and institutional thickness of places. Places with a longstanding specialization in well-known agri-food products equipped with a support system of local institutions have the ability to underpin the actors in SFSCs. Conversely, places with a rather poor reputation and lack of institutions can prove to be hostile. The differentiated potential of the Geographical Indication (GI) cheeses, Parmigiano-Reggiano in Parma and Feta in Thessaly, illustrate the point.

Regional variations/similarities

The discussion on SFSCs highlights the fact that notable regional variations result in SFSCs with drastically differentiated attributes and dynamics.

The preceding chapters testify to the selective character of SFSCs in high-income countries. In Ontario, the owners of small agri-food firms are highly motivated towards sustainable and local production (Chapter 5). Perrin *et al.* (Chapter 10) found a strong association between the establishment of new open-air markets and the income of the resident population in Montpelier peri-urban areas. Kalfagianni (Chapter 9) has shown that alternative retailers operating in the Netherlands prioritize complementary aspects of food quality at the expense of price.

The picture is radically different in the crisis-hit countries of Mediterranean Europe where employment creation, affordable prices and local development are found to be centre stage. The urgent need for the fulfilment of these criteria is, too often, realized at the expense of core criteria – local origin, environmental safety and nutritional content of products. Both Skordili (Chapter 6) and Anthopoulou and Goussios (Chapter 8) found that in crisis-hit Greece, the predominance of criteria such as cheap price and patriotic consumption have led, in several instances, to the gradual replacement of local food by the vague indication "food produced within the national territory". Hence, in difficult times, there is a drastic draw-back of the local.

However, the chapters in this book have revealed profound similarities as well. The majority of the actors in SFSCs face a common set of problems all over the world. Too often, we make a generic reference to private firms or enterprises. The use of this broad term does not always make clear that the vast majority of actors in SFSCs are micro or nano firms, extremely vulnerable to the many threats and instabilities of the wider business environment. It is indicative that even in the case of Canada, the annual sales figure – for a percentage as high as 37 per cent of the firms surveyed – was less than $50,000. Blay-Palmer *et al.* (Chapter 5) are right to observe that these figures point to financial precariousness.

The number one threat for these fragile small-scale actors is fierce competition from bigger, conventional competitors. There are several examples of appropriation of successful local products and business ideas by larger firms. The examples of Parmigiano-Reggiano in Italy and craft beers in Greece illustrate the point. More importantly, conventional competitors have embraced the two most successful brands in the food industry: local and organic products. Anthopoulou and Goussios (Chapter 8) show that bigger competitors co-opt valuable intangible assets as well – the symbolic and material heritage of the long-standing cheese production system in Thessaly, Greece.

A second major common risk for micro agri-food firms operating in strikingly differentiated business environments, such as affluent Ontario in Canada and crisis-hit Greece, is the lack of access to finance. Large numbers of small-scale local food initiatives are short-lived and have been thrown out of market due to the lack of capital and fierce competition. Arfini and Mancini (Chapter 7) reveal the gradual withdrawal of smaller farmers and dairies from the Parmigiano-Reggiano production system and their replacement by larger firms. Another significant consequence of the lack of finance, as shown by Blay-Palmer *et al.* (Chapter 5) and Skordili (Chapter 6), is the prevention of innovation, delays, or even cancellations, of significant arrangements envisaged in their initial business plans.

Finally, a third set of similarities is found in the limited networking capacities observed across different geographical settings. There is wide agreement that the formation of local networks can overcome the scale problem of small actors in the supply chain. Several chapters document the inability to form networks among producers, consumers and other stakeholders. Blay-Palmer *et al.* (Chapter 5) acknowledge the inability of small agri-food firms to engage with the local community and build relationships with various sectors. Anthopoulou and Goussios (Chapter 8) observe a lack of cooperation due to the family structure of firms. In the case of Parmigiano-Reggiano the thick institutional system of the region connects and coordinates several stakeholders in dense networks.

Governance

Agency, the capacity to act and to constitute in the context of plurality and diversity, and fairness were the central aspects of governance examined in detail in this book. In this context, a common theme that emerges is that food is – and should be - regarded as a political object. Food, essential for human survival, is not only a commodity that is produced and sold on the market. Rather, it is an object of social struggle, of demands for access to health and nutrition, of the right to cultivate while respecting boundaries and with an ethos of stewardship towards the planet, and of the right to sovereignty, particularly for the most vulnerable and marginalized. As Morgan and Santo note (Chapter 2), in order to re-moralize food and, in fact, in order to be able to have any moral and ethical discussion around food, we need to consider it as a political object.

To a large extent, SFSC initiatives do exactly that. As Grasseni notes (Chapter 3), SFSCs in the form of collective food procurement provide a key outlet for citizens/consumers in taking responsibility for the global food system. Other contributors too – Partzsch (Chapter 4), Blay Palmer *et al.* (Chapter 5), Peuker (Chapter 11) and Perrin *et al.* (Chapter 10) – underline the significance of such initiatives for collective innovation and collective action that can galvanize common concerns about intensive and commodified forms of production, distribution and consumption of food. Yet, these chapters also show that the relation between SFSCs and political objectives around food is not automatic. Indeed, engagement with such initiatives can also be motivated by investment opportunities and broader trends towards "healthier" and "alternative" lifestyles, as well as the desire, among relatively wealthy consumers, for a plurality of choices.

The form of political agency that emerges from these initiatives also needs to be scrutinized with caution. A central critique of SFSC initiatives is their rather elitist and exclusive nature which tends to deliver agency to those who already – to a certain extent – have it. Both Partzsch (Chapter 4) and Blay-Palmer *et al.* (Chapter 5) note that access to land, a key resource for the flourishing of such initiatives, is extremely difficult due to high land prices. Accordingly, agency is partly determined by capital availability (see also Peuker, Chapter 11). Likewise, Anthopoulou and Goussios (Chapter 8) issue caution about too easily assuming agency in respect of the most vulnerable. In their discussion on GI schemes under the national Protected Designation of Origin (PDO) regulation for Greek Feta cheese, Anthopoulou and Goussios point out that these schemes tend to favour the larger producers. Furthermore, as Partzsch notes, agency needs to be accompanied by accountability as without the latter it can be exercised by the more powerful against the weakest with limited or no consequences.

Skordili (Chapter 6) provides a more optimistic view of agency, observing the importance of SFSC initiatives in safeguarding employment in crisis stricken places and providing access to the products of vulnerable local producers excluded from conventional food systems. Morgan and Santo (Chapter 2), in turn, emphasize the need to think about collective agency beyond individual initiatives. To be able to become transformative SFSCs in their multitude need to embrace diversity and form a common identity that will enable them to become a true alternative to powerful and more homogenized actors in globalized supply chains.

Finally, fairness and justice are key concerns of SFSC initiatives. In the past, as this book notes, these demands were marginalized in favour of environmental concerns. However, eclectic demands of environmental sustainability have resulted in the further marginalization of vulnerable, resource-weak groups (see Morgan and Santo, Chapter 2). By introducing fairness and justice as key values, SFSC initiatives have greater promise to simultaneously foster environmental sustainability and social justice in the agri-food system. In this context, however, and as this book demonstrates, results from current efforts are mixed.

On the one hand, Grasseni (Chapter 3) underlines the importance of collective food procurement initiatives in advancing fair prices and moral solidarity among participants. She further notes the possibility of scaling up when local networks link up with Fairtrade shops built around social justice values beyond the case of food. Partzsch (Chapter 4) points out that fair prices for producers are possible in SFSC initiatives when these promote transparency, particularly in terms of accounting. Skordili (Chapter 6) discusses the multiple dimensions of fairness that are fostered through micro-entrepreneurship initiatives including inclusive and equitable development and fair access to food, as well as fair labour practices and fair prices for producers. Finally, Arfini and Mancini (Chapter 7) note the conditions for fair competition provided by short food supply initiatives in contrast to their long supply chain counterparts.

On the other hand, Anthopoulou and Goussios (Chapter 8) demonstrate that despite good intentions GI systems failed to provide a fair redistribution of value in the supply chain, particularly for more vulnerable producers. Perrin *et al.* (Chapter 10) bring to our attention the possible trade-offs in terms of fair value when different stakeholder groups are considered, particularly farmers, consumers and retailers. Finally, Kalfagianni (Chapter 9) highlights that alternative food retailers address many of the fairness concerns associated with long food chains in the material, social, political and environmental dimensions, yet they are unable to constitute an adequate response by themselves to the conventional food regime.

In short, this book demonstrates that alternatives to the dominant intensive and hugely concentrated agri-food system are possible and thriving. Yet, a concerted effort to move beyond their silos, create links with one another and speak with a common voice while recognizing and respecting diversity is a necessary condition for realizing their full potential.

Index

Page numbers in **bold** denote tables, those in *italics* denote figures.

For Product Safety Concerns and Information please contact our EU representative GPSR@taylorandfrancis.com
Taylor & Francis Verlag GmbH, Kaufingerstraße 24, 80331 München, Germany

www.ingramcontent.com/pod-product-compliance
Ingram Content Group UK Ltd.
Pitfield, Milton Keynes, MK11 3LW, UK
UKHW022128190926
475702UK00013B/1698

* 9 7 8 0 3 6 7 5 8 2 9 9 9 *